Geology of Turkey

R. Brinkmann

ELSEVIER SCIENTIFIC PUBLISHING COMPANY
Amsterdam — Oxford — New York 1976

The distribution of this book is being handled by the following publishers:

for the German Federal Republic, the German Democratic Republic, Austria, Schwitzerland, Lichtenstein, East Europe, China, Democratic People's Republic of Korea, Cuba, the Democratic Republic of Vietnam and of Mongolia

Ferdinand Enke Verlag
Stuttgart

for the U.S.A. and Canada

American Elsevier Publishing Company, Inc.
52 Vanderbilt Avenue
New York, New York 10017

for all remaining areas

Elsevier Scientific Publishing Company
335 Jan van Galenstraat
P.O. Box 211, Amsterdam, The Netherlands

Library of Congress Catalog Card Number:

ISBN Ferdinand Enke Verlag 3-432-88271-8
ISBN Elsevier 0-444-99833-0

Printed in Germany

Introduction

Today, the Mediterranean area is a focal point of geological research; however, most investigations do not take its eastern boundaries into account. *Geology of Turkey,* therefore, would like to bridge this gap. This book is not meant to be a handbook; rather, it is designed to offer a brief introduction to the geology of Turkey, its structure and geotectonic problems as well as a guide to Turkish geologic literature. The paleogeographic and tectonic relations to the neighboring areas are shown, and some references to the paleontology and ore deposits are included.

Turkey has been loosely subdivided into sections. In particular, the inner highlands between the border mountains (i. e., approximately the area between Izmir, Denizli and Bursa, Bolu and Konya, Sivas and Niğde, Erzincan-Erzurum-Kars and Malatya-Elaziğ-Van) are referred to as „Middle Anatolia“.

The description is quite condensed. Many details are recorded in maps and tables. Thus, space has been allocated for historical observations and – above all – for numerous citations. Older publications have not been included in the references; they can easily be found in the summaries, beginning with chapter 2.

This book is the result of eight years of teaching and scientific research at Ege University at Izmir. Its publication gives me an opportunity to thank my colleagues in Turkey and Europe, and especially my co-workers at the Izmir Geological Institute, for their support. I am indebted to Mrs. Irene Woodall for the translation, to Dr. E. T. Degens for a final revision of the manuscript, to Mrs. H. B. Müller, Hamburg, and to Mr. E. Vural, Izmir, for the art work.

R. Brinkmann

Contents

Part One Introduction

Chapter 1 The History of Geology in Turkey

Turkey, together with the other countries in the Near East, is the cradle of geology and mining. Technical utilization and scientific knowledge of rocks has come primarily from those areas, and in particular from Anatolia, where many fundamental geologic observations have emanated.

In prehistoric times, man made his tools primarily from rock. Clay was employed later; burnt-clay vessels are known in Asia Minor since 6500 B. C. But other minerals as well have been used as early as in the seventh millenium. Malachite, azurite, cinnabar, and ochre served as pigments (Fig. 1). Occasionally, utensils were made of lead and copper.

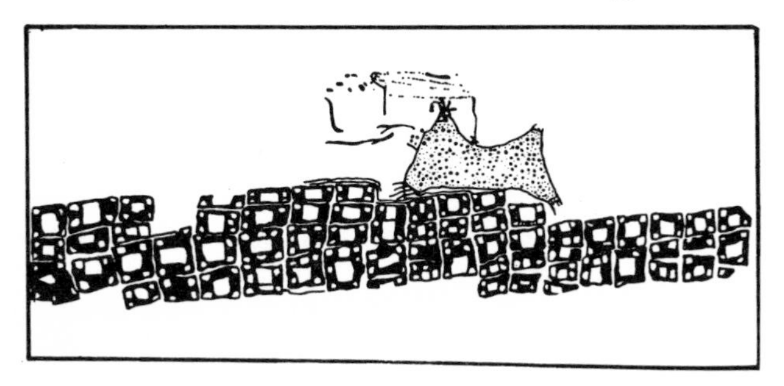

Fig. 1 Wall painting, about 6600 B. C., Catalhüyük near Konya, interpreted as a view of the neolithic settlement with an active volcano of the Hasan Dag in the background (after *J. Mellaart,* Catal Hüyük, A neolithic town in Anatolia. Thames and Hudson, London 1967, pl. 59–60).

Gold however, is the oldest of all metals. Already in Neolithic times, it has been worked into ornaments. Anatolia was famous for its gold, many legends tell about it. King Midas whose touch turned everything to gold, is said to have ruled there. In recorded times, there was King Kroisos of Lydia. His proverbial wealth was based on placer deposits in western Anatolia. At approximately 650 B. C. he had the first coins made from this gold.

Silver, in its native state, is rarer; however, as early as in the third millenium, man tried to smelt it from sulphide ores – evidence that even at that time mining operations had penetrated to greater depths. East Anatolia has always been known to be rich in silver. Hattusas, the capital of the Hittites, means „silver-town". The smelting of less precious metals required more technical know-how. The oldest copper mines probably were located in Egypt. About the middle of the third millenium, Cyprus which gave its name to the metal, and East Anatolia became major copper mining areas. Occasionally, iron had been in use since the middle of the second millenium B. C., however, it did not spread across the Old World until 1400 B. C. It was a slow process; as a matter of fact, the Trojan War, in the 13th century, was still fought with bronze weapons (*Forbes,* 1963/1966).

The Greeks, who, at the beginning of the first millenium, populated the shores of Anatolia, and in particular the Ionian area (western coast around Izmir) were more conscious of the phenomena of nature than the peoples of the Ancient Orient (*Berger,* 1903; *Adams,* 1954). The philosopher Thales, who lived in Miletos (south of Izmir) at around 600 B. C., raised the question of what earth consisted of. Thus, he became the founder of earth science as we understand it today. Being a citizen of a town which was situated on a peninsula in a sweeping bay, he regarded the mainland as a disc floating on top of the sea. If the disc sways, it can be felt in the form of earthquakes. His student, Anaximander, was the first one to design a cosmogony based on science. Earth has gradually dried out, and the salty sea contains the rest of the waters. In the humid mud, fish appeared first which gradually turned into man. According to Anaxagoras of Klazomenai (Gulf of Izmir), earth and the other planets have sprung from a primeval nebula. He used a meteorite fallen in 468 B. C. to prove the material unity of the solar system. About the middle of the fourth century B. C., Xenophanes of Kolophon (south of Izmir) and Eudoxos of Knidos (on the Marmaris Peninsula) interpreted salt lakes and marine fossils in central Anatolia as remnants of a former sea. Not much later, during his travels, Herodotos of Halikarnassos (southwestern Anatolia) collected observations on the present changes of morphology. He regarded the valleys of Miletos and Ephesos as well as the delta of the Nile as alluvial river deposits. Around 350 B. C., Aristotle spent some years in Anatolia; he was of the opinion that the geological changes, such as uplifting, subsidence, shifting of the coastlines, and climatic changes occurred unnoticeably slow and in an ordered manner. He compared the mountain ranges to sponges; at their base the groundwater gushes forth in the form of springs (*Gohlke,* 1955). His student, Theophrastos of Lesbos, gave an up-to-date description of the mineralogic and mining knowledge of his time in his book, *About Rocks* (*Eichholz,* 1965). During his travels, Strabo of Amaseia (Amasya, 65 B. C. – 23 A. D.) became convinced that the shifting of the coastline was caused by vertical movements of the earth's crust rather than by alluviation. Shrewdly he perceived that the Katakekaumene area (east of Izmir, p. 81) was an extinct volcanic field. He regarded the volcanoes as safety valves for the compressed steam in the earth's interior. When the steam escapes, then the earth begins to quake (*Serbin,* 1893).

During the first centuries A. D., minerals were exploited at numerous places in Anatolia. Many of the mines that had been in operation then, are still being operated today. Coal and fossil oil were of minor importance during antiquity. However, oil seepages were known in southeast Anatolia. The natural gas flame of Chimaera (south of Antalya) has been burning for more than two and one-half millenia.

The founding of towns in Asia Minor necessitated a planned usage of the groundwater as early as in the third millenium B. C. The legend of the rivers of Hades shows that the ancient peoples were familiar with karst hydrology. During Hellenistic-Roman times, aqueducts and cisterns were constructed in the then metropolitan cities of Byzantium, Smyrna, Pergamon, and Ephesos, partly with pressurized pipes.

Antiquity ended with a decline of science. A change did not occur until the spreading of the Islam. It brought science to new heights during the years of 800 to 1300 A. D., aided by the establishment of a large empire. The Islamic religion sees nature as an entity on a religious foundation. The objects of nature represent random phenomena within the framework of a general, timeless order. Only the „hakîm", to whom wisdom means more than science, will obtain full knowledge. The great sages of that epoch were al-Masudi, al-Biruni, and ibn Sina (between 950 and 1050 A. D.). According to them, the mountains, as we see them today, have been eroded to their present shape. Their debris is carried to the sea and thus is being gradually fragmented. Sand and mud are deposited at the bottom of the sea in chronological order. Periodically, the mountain ranges are uplifted by seismic forces, the dry mud transforms into rock, and the enclosed animal remnants become fossilized. Finally, the solidified sediments enter into the same cycle (*Nasr,* 1968).

The Islamic way of weltanschauung enabled the scientists of those times to take over the inheritance from antiquity and to disseminate it in Europe together with the results of their own investigations. In Europe, however, the further development took a drastic turn: the natural sciences – as we know them today – were born. Contrary to Islamic thinking, they do not start with the unity of nature but with individual objects in nature. Observations are primary, and they are being intertwined with causal laws. The religious element was lost, and the scientist and specialist took the place of the sage. The controversy between the two ideas of science paralyzed the intellectual life in Asia Minor for half a millenium.

A first change came about with the founding of the Osmanian state. The sultans had the works of ancient geographers translated and had them supplemented by Turkish travel descriptions. Medicine was another science of practical implications. Since ancient times, it had been related to geology by the medicinal belief in certain rocks, soils, and waters. During the last centuries in Turkey, therefore, geography and medicine were the cradle of all natural sciences, including geology.

The country remained closed to foreigners for a long time. Only reluctantly did its borders open in the 19th century for the first European scientists, like *Ainsworth, Forbes, Hamilton, Hommaire de Hell, Russegger, Spratt, Strickland,* and *Texier.* Their descriptions which principally deal with the general knowledge of the country and its archaeology, contain also quite a few geological observations. With his book, *Asie Mineure* (1853–1869), which is fundamental for Anatolia, *De Tchihatcheff* initiated the age of geologic travel descriptions. His predecessors in the European part of Turkey were *Boué* (1840) and *Viquesnel* (1868). In Anatolia he was followed by *Abich* (1878–1887), *Naumann* (1893), *Schaffer* (1900–1918), *Grothe* (1911–1912), *Philippson* (1910–1915), and *Chaput* (1936). At the end of the 19th century, specific geological investigations became more prevalent. At first, mainly Austrian geologists entered the country: *Bittner, Von Bukowski, Hoernes, Kober, Neumayr, Tietze* and *Toula.* During the first decades of this century, the scientists came mainly from Germany: *Berg, Frech, Kossmat, Paeckelmann* and *W. Penck.* In addition to these foreigners, Turkish scientists began to take part in the investigations.

The activities of these geologists prepared the ground for the further development of earth sciences in Turkey. The University of Istanbul, which was founded in 1915, asked *W. Penck* to be its first professor of geology; he was succeeded by *Pamir. Paréjas* and *Chaput* were occasional guest lecturers. A geological survey, the Maden Tetkik ve Arama Enstitüsü, was established at Ankara in 1935. *Akartuna, Altınlı, Arni, Baykal, Bayramgıl, Blumenthal, Dacı-Dizer, Egeran, C. Erentöz, L. Erünal-Erentöz, Ketin, Kleinsorge, Kovenko, Lahn-Ilhan, Lokman, Okay, Ortynski, Pamir, Patijn, Salomon-Calvi, Stchépinsky, C. E. Tasman, Ternek, Tokay, Tolun, Tromp, Ünsalaner-Kırağlı, Van Der Kaaden, Van Wijkerslooth, Yalcınlar,* and many others attempted the first geological mapping of the country. Today, geology is taught at all universities (Adana, Ankara, Erzurum, Istanbul, Izmir, and Trabzon). At the Geological Survey, a staff of 700 scientists is continuing the work of the pioneer generation in all branches of pure and applied geology.

Chapter 2 The Geological Literature of Turkey

The following publications report on the geology of Turkey:

1. The Maden Tetkik ve Arama Enstitüsü (MTA) is publishing:
   a) monographs concerning regional geology, paleontology, and economic geology (bilingual or in foreign languages) under the heading of *Publications de l'Etudes et de Recherches Minières* (MTA Yayınları). The series A–D (1937–1958) have been completed. The new series starts with no. 101 (1958).
   b) short communications in the *Bulletin of the Mineral Research and Exploration Institute of Turkey* (foreign languages) starting with no. 46 (1954), or in the MTA *Dergisi* (Turkish). The predecessor of both series is the MTA *Mecmuası,* nos. 1–45 (1936–1953, bilingual).
   c) in addition, the MTA has extensive archives consisting of unpublished reports and maps which are occasionally cited in this book.

2. The universities also publish scientific reports. The most important ones are the *Revue de la Faculté des Sciences de l'Université d'Istanbul (Sc. Nat.)*, which appears in foreign languages, and the *Monografileri* of the same faculty (Turkish).

3. The *Bulletin of the Geological Society of Turkey* (foreign languages and Turkish), starting with volume 1 (1947).

4. In addition to the series listed above, foreign scientific journals, in particular Austrian, French, German, Italian and Swiss journals, contain many informations about the geology of Turkey.

The first geological maps of Turkey have been done by *De Tchihatcheff* (1867) and *Frech* (1916). The MTA has combined its work in one map at a scale of 1:500,000 (18 sheets, 1961–1964).

Comprehensive studies on the geology of Turkey have been presented by *Philippson* (1918) for central and western Anatolia; by *Oswald* (1912) for eastern Anatolia; by *Egeran* and *Lahn* (1948), and *Ten Dam* and *Tolun* (1962) for the entire country. *Pamir* (1960) and *Baykal* (1971) have focussed on stratigraphic problems. A volume edited by *Campbell* (1971) offers the most recent summary on Turkey.

Stchépinsky (1946, 1947) has reported on the fossil fauna and flora of Turkey.

Studies of major parts of Turkey have been done by *Philippson* (1910–1915) and *Brinkmann* (1971b) for West Anatolia, by *Kopp* et al. (1969) for Thracia, by *Brunn* et al. (1972a) for the western Taurus Mountains, by *Blumenthal* (1960–1963) for the Taurus and Antitaurus, by *Altınlı* (1966) for eastern and southeastern Anatolia, and by *Tolun* (1962) for southeastern Anatolia.

A complete list of the geological literature of Turkey is in preparation (*Brinkmann* and *Erol*). At present, the most comprehensive index is by *Furon* (1953). For references on older papers see *Toula* (1904), *Von Bukowski* (1904), *Oswald* (1912), and *Philippson* (1918). Recent summaries are cited at the beginning of chapters 3 to 21. The titles of papers which have been published most recently are available in a Turkish list (Turdok).

Part Two Historical Geology

Chapter 3 Crystalline Basement Rocks

The areal extent of regionally metamorphosed rocks in Turkey is fairly well understood; however – except for some limited areas – no modern studies have been undertaken. Recent summaries on the metamorphic rocks have been prepared by *Van Der Kaaden* (1971) and *Brinkmann* (1971a).

In Turkey, crystalline rocks are found in four major areas (Fig. 2):

1. the Istranca Massif (*Hochstetter*, 1870) in Thracia;
2. the Menderes Massif (Lydian-Carian Massif, *Philippson*, 1905) in southwestern Anatolia;
3. the Kırsehir Massif (Kızılırmak Massif, *Chaput*, 1936) in central Anatolia; and
4. the Bitlis Massif (*Arni*, 1939a) in southeastern Anatolia.

In addition there are numerous smaller crystalline massifs, some of which are described below.

Istranca Massif

This mountain range has not been investigated since *Pamir* and *Baykal* (1947). According to their studies, the southwestern part is composed of highly metamorphosed rocks and the northeastern part of slightly metamorphosed rocks. The northwest-trending schistosity is parallel to the longitudinal axis of the mountain range.

The banded Fatmakaya paragneiss masses are the oldest metamorphic rocks. They are intruded by coarse-grained, migmatized Kırklareli augen gneiss.

The gneiss approaches within 20 km the Paleozoic of Istanbul, however, no petrographic transition between the two complexes is discernible. From this we can infer that the gneiss is older than the Paleozoic, i. e. pre-Ordovician.

The epimetamorphic sequence consists of dark phyllite, sandy calcareous phyllite, and light, fine-grained marble. It rests on the gneiss unconformably and with a hiatus in metamorphism (unpubl. data). There are traces of Mesozoic fossils. In the vicinity of the Black Sea, it is covered by unmetamorphosed Upper Cretaceous sediments.

The Istranca Massif is part of the Rila-Rhodope Massif. The same rocks are probably underlying major parts of the Thracian Tertiary basin as the *Macedonian-Thracian plate* (*Ivanov* and *Kopp*, 1969b: 118). Although so far no exact parallels exist, the Istranca gneiss corresponds to the oldest supposedly Cryptozoic crystalline rocks of the Rhodope range. The epimetamorphic sequence continues into southern Bulgaria as the Sakar-Strandca series (p. 95). Here it consists of Triassic and Jurassic rocks which habe been metamorphosed prior to the Upper Cretaceous (*Bojadjiev*, 1971: 505).

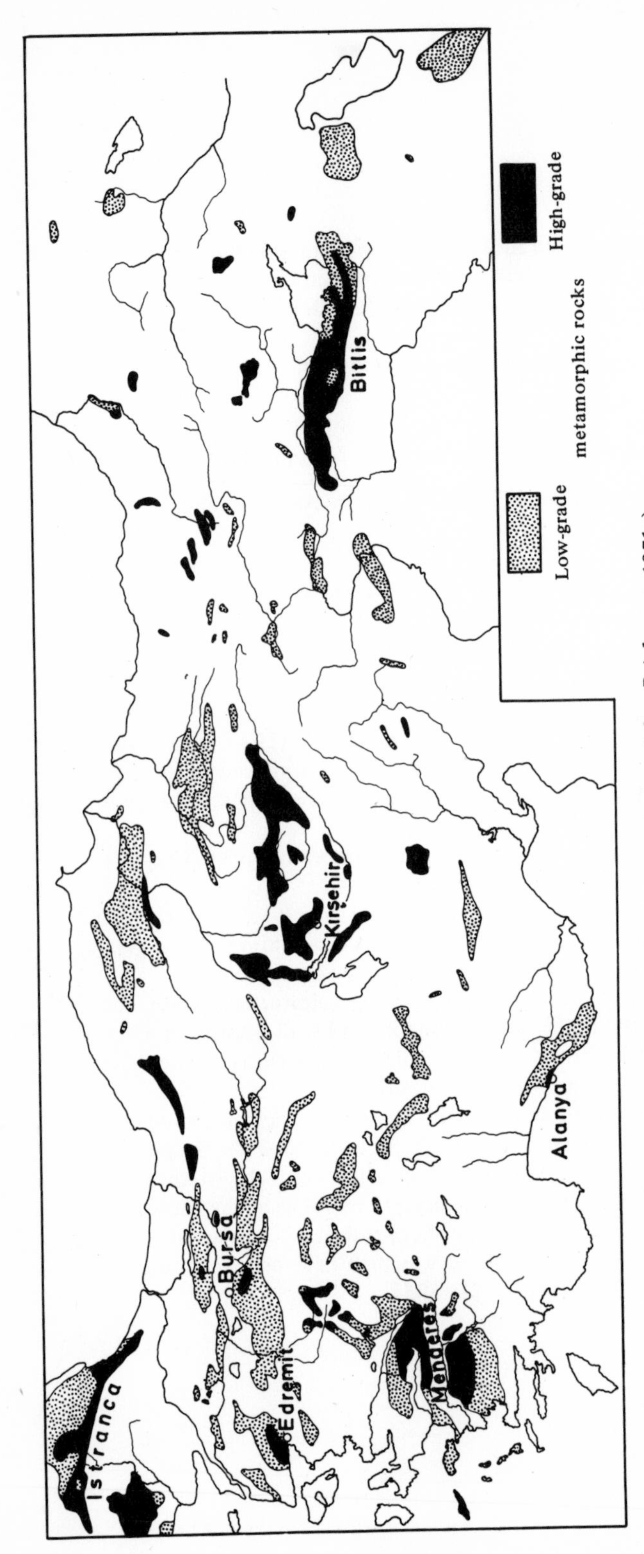

Fig. 2 Distribution of metamorphic rocks in Turkey, irrespective of geological age (after *Brinkmann*, 1971a)

Menderes Massif (Fig. 3)
The western and best known part of the Menderes Massif consists of several dome-shaped uplifts. The biggest one of those uplifts extends on either side of the Büyük Menderes river. North of that area, there are smaller uplifts along the Küçük Menderes river, the upper Gediz river, and in the Eğrigöz-Karakoca mountains. The core of these domes is formed by augen gneiss. It is surrounded by rocks which in the lower levels consist mostly of mica schist, phyllite, and metaquartzite, in the upper levels mainly of marble. Along the southern slope of the massif the marble contains lenses of diaspore-hematite or corundum-magnetite, which best can be interpreted as metamorphosed bauxite deposits (*Önay,* 1949).

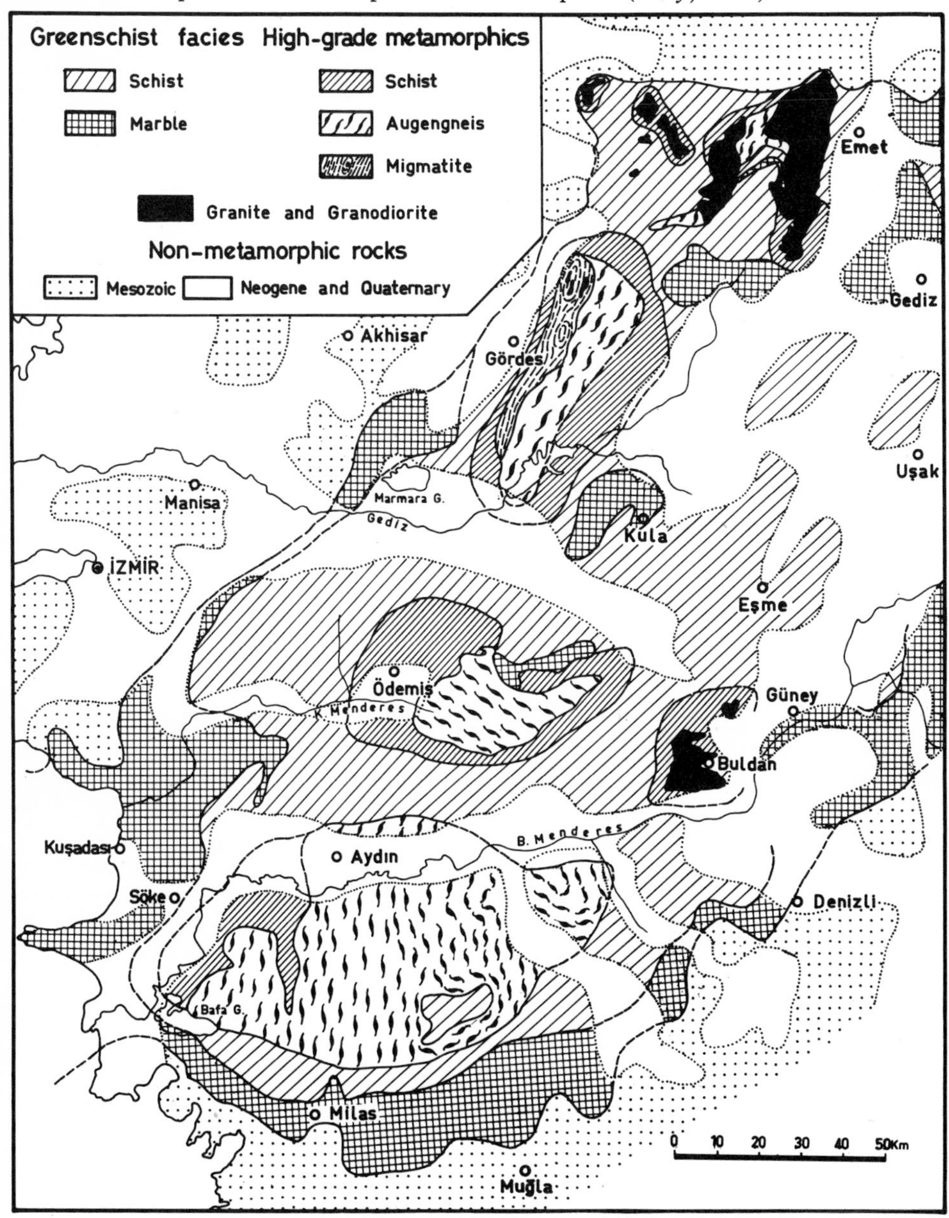

Fig. 3 Crystalline rocks and their metamorphic facies in the Menderes Massif, southwestern Anatolia (after *Schuiling* 1962, *Izdar* 1971, mscr., *Dora* 1972, mscr. and others)

In the center of the domes, the augen gneiss is coarsely crystalline and even massive; towards the borders it takes on a more and more schistose and fine-grained texture. In the southern (*Schuiling,* 1962: 71) as well as in the northern part (*Dora,* 1969) of the Menderes Massif the foliation planes of the gneiss and the schist synclines within the gneiss have a NNE direction. This structural trend especially in the southern core strongly contrasts with the periclinal dip in the schist mantle. *Schuiling* (1962: 74) and *De Graciansky* (1966: 299) interpreted the observed relationship as an unconformity. The augen gneiss has inherited its foliation from a previously folded sequence and subsequently was covered by the schists of the mantle.

In general, the augen gneiss is regarded as an anatexite. According to *Schuiling* (1962: 75), *Başarır* (1970: 29), and *Dora* (1969: 17), it originated from sediments, and according to *De Graciansky* (1966: 304) and *Izdar* (1971: 500) it originated principally from magmatic rocks. In places, the augen gneiss is intrusive in the schist mantle. The contact is sharp and often characterized by tourmalinization. Blocks of schist are floating in the gneiss, and gneiss veins have penetrated the schist. In other places, a gradual transition from schist to gneiss caused by an increasing feldspathization (*Başarır,* 1970: 29 f.) is developed.

Contrary to these relationships is the observation that the schists surrounding the gneissic cores consist prevalently of metamorphic rocks in green-schist facies. The schist enveloping the southern gneissic core at its contact belongs to the quartz-albite-epidote-almandite subfacies. Going outward, it is overlain by the lower metamorphic zones of the Barrow sequence (*De Graciansky,* 1966; *Başarır,* 1970). Schistose rocks in amphibolite facies are rare. They occupy synclines in the southern gneissic core (*Schuiling,* 1962: 78) and form narrow seams around the middle (*Izdar,* 1971) and northern (*Dora,* 1972: 136) gneissic cores.

Scotford (1969) tried to resolve this problem in assuming that the augen gneiss and the feldspathized schist are the product of an alkali metasomatosis at 400° to 550° C. This theory is not convincing in view of the geological occurence of the gneiss. In contrast, *Dora* (1972) and *Izdar* (1971) assume two metamorphic phases. During the earlier phase, the metamorphism in the massif advanced to the amphibolite isograde, and in the center to anatexis. During the later phase, the massif was metamorphosed in a retrograde order. In the highlands of Eğrigöz-Karakoca the earlier phase produced migmatite, whereas during the later phase granitic rocks were formed (*Öztunalı,* 1973: 104; *Uz* 1973: 165). According to *Dora* (1969, 1972) the first metamorphic phase reached temperatures of up to 680° C, and the second one of 500° to 550° C.

The age of the crystalline rocks and the time they became metamorphosed can be estimated by the following geological criteria.

Schuiling placed the unconformity between core and mantle in Late Caledonian times. However, that phase is of little importance in Turkey (p. 22). A previous, Early Caledonian or Assyntian tectogenesis seems much more probable. Thus, the age of the material from which the augen gneiss is derived, must be placed in Cambrian or Precambrian time. Near their base the schist series contains lenses of former sedimentary iron ore which, in other parts of Anatolia, is characteristic for the Ordovician or Devonian period. *Yalcınlar* (1963: 15) mentions crinoid-bearing phyllitic rocks which also could be of Devonian origin. The base of the marble zone at Muğla is dated by fossils, questionable ones of Lower Carboniferous age, certain ones of Permian age (*Önay,* 1949: 364; *Metz,* 1955: 265). The top of the marble zone in southwest Anatolia is of Late Cretaceous age (*Dürr,* pers. comm.) on the basis of rudist findings. A slightly earlier metamorphic phase is indicated at Bodrum by Triassic and Liassic recrystallized limestones overlain by unmetamorphosed limestone of uppermost Jurassic age (*Brinkmann,* 1967: 5). However, *De Graciansky* places this area into the Taurids (1972: 539). On the northwestern border of the massif, non-metamorphosed Mesozoic rocks are transgressive on the crystalline basement, mostly Upper Cretaceous rocks (*Akartuna,* 1962a: 14; *Oğuz,* 1967; *Ayan,* 1973: 90). The northern slope of the Eğrigöz-Karakoca mountains is structured in a similar way (*Dora*, 1969: 16); whereas *Kaya* (1972b: 494) found two metamorphic hiatuses: the older one between the Menderes crystalline complex and probably Jurassic strata, and the younger one between these and the Upper Cretaceous beds.

The following radiometric dates are available for the Menderes Massif:

	Total rock Rb/Sr	Orthoclase Rb/Sr	Muscovite Rb/Sr	Biotite Rb/Sr	Zircon U + Th/Pb	
Augen gneiss, southern core	490 ∓90 529		66 ∓4	22,2 ∓1,3		*Jäger*, pers. comm. *Schuiling*, 1973: 44
Augen gneiss, middle core				20,2 ∓3,7		*Jäger*, pers. comm.
Egrigöz granite, northern core	167 ∓14	31 ∓5		29 ∓3	 69,6 ∓7	*Öztunalı*, 1973: 1 *Bürküt*, 1966: 136

A uraninite vein in the southern gneiss core was dated at 268 ∓ 60 m. y. (*Durand*, 1962). It is uncertain whether the Alacam granite (p. 38) belongs to the Menderes Massif.

The Menderes Massif, thus, represents a crystalline complex, its core rocks being of lowest Paleozoic or of Cryptozoic age. The schists of the mantle are derived from Lower Paleozoic, and the marbles from Upper Paleozoic and Mesozoic sediments. The massif was subjected to several metamorphic events. The one prior to the last metamorphism, probably a Jurassic phase, affected the entire massif, and in its core advanced to anatexis. The last, post-Cretaceous phase was on the whole retrograde.

All these results refer to the western part of the massif. The eastern part has so far been little studied. Thus, it is questionable, whether the Sultan Dağ should be included in the Menderes Massif (*Brinkmann*, 1971a: 889). Here, metamorphism as well as folding are pre-Triassic (*Haude*, 1972: 419; *Brunn* et al., 1972a: 520). Such a metamorphic hiatus coupled with a Late Variscan tectogenesis has never been shown for the western part of the massif. Only *Yalcınlar* (1963: 19) and *Akarsu* (1969: 4) mention slightly marblized Permo-Carboniferous (?) limestones unconformably overlying metamorphic rocks of the schist mantle at Baba Dağ near Denizli.

Kırşehir Massif

This massif is not very obvious on the map as a geological unit. Large areas are covered by younger sediments and only parts of the highly crystallized core are visible. In the northwestern part of the massif, *Ayan* (1963: 240) describes paragneisses of amphibolite facies: sillimanite and biotite gneiss, amphibolite, metaquartzite, and marble with northeast-trending schistosity. In the northeastern part the sequence of the massif is divided into two series; according to *Pollak* (1958) and *Vaché* (1963) biotite mica schist with quartzite and marble are unconformably overlying hornblende- and sillimanite-bearing muscovite-biotite gneiss.

Both series show west-southwest-trending folds which are verging towards the south. Granitic rocks occur as anticlinal plutons. In the southern part of the massif near Niğde, there are thick marble formations accompanied by gneiss in amphibolite facies (*Metz*, 1939b: 315; *Blumenthal*, 1941: 53).

Unmetamorphosed surface rocks of Paleogene and Uppermost Cretaceous age are widespread. *Ketin* (1966a: 28) thus draws the conclusion that the massif was formed during the Laramian phase. Indeed there are granitic rocks of that age in the massif (*Ayan*, 1963: 324).

The crystalline schists, however, are thought to be considerably older according to some exposures on the southeastern slope of the massif. At Bünyan (*Baykal,* 1945: 136) and at Yahyalı (unpub. data), east and southeast of Kayseri, unmetamorphosed Devonian sediments are transgressive on phyllitic rocks. If the latter are part of the mantle of the Kırşehir gneiss, then at least an Early Paleozoic age for the metamorphism would be established.

Leuchs (1943: 36) suspected a huge old block in the subsurface of middle Anatolia, the *Inner Anatolian Massif.* Its higher lying eastern part is the Kırşehir Massif. Its deeply submerged western part, the *Lycaonian Massif,* forms the base of the Tuz Gölü basin south of Ankara. Paleogeographic data, indeed, argue for the probability of an inner Anatolian island during Mesozoic times which partly consisted of crystalline rocks (p. 50). This supports the idea of a pre-Mesozoic regional metamorphism of the Kırşehir Massif.

Bitlis Massif

According to *Ternek* (1953b: 4), *Tolun* (1953: 81; 1962: 211), *O. Yılmaz* (1971: 24) and *Pişkin* (1972: 8) the oldest crystalline rocks of the massif are partly of magmatic origin: amphibolite, hornblende gneiss, and light-colored gneiss; and partly of sedimentary origin: sillimanite-bearing foliated gneiss and muscovite-biotite schist. They all belong to the amphibolite facies, and their schistosity is generally northwest-trending. Granitic and pegmatitic rocks have intruded. These basement rocks are overlain – with angular unconformity and metamorphic hiatus – by greenschists with quartzite, and finally by marble. On the southern slope of the massif the latter turns into limestones of Permian and possibly Mesozoic age.

The geological and petrographical evidence for at least two metamorphic phases is confirmed by radiometric measurements (*O. Yılmaz,* 1971: 205). Ages given in m. y. are:

	Total rock (Rb/Sr)	Zircon (U + Th/Pb)	Feldspar (Rb/Sr)	Chloritized biotite (Rb/Sr)
	920 ∓ 224			
	596 ∓ 88			
Basement	519 ∓ 232			
metamorphites	505 ∓ 37			
	427 ∓ 54			66
	325 ∓ 3	639 ∓ 15	120 ∓ 10	97 ∓ 8
Basement granite	324 ∓ 44			
	351			

According to these measurements, the basement rocks are of Late Proterozoic age. They probably belong to the same old crystalline mass which is known to underly Cambrian strata in neighboring Iran (*Stöcklin* et al., 1964: 8). In the Bitlis Massif, however, it was subjected to a Caledonian recrystallization. The granite intruded in Early Variscan time. The rocks of the mantle, which are resting transgressively on the base, owe their appearance to a Late Mesozoic metamorphism which extended far beyond the massif itself (p. 96).

Kaz Dağ Massif

In recent times, the Kaz Dağ has been studied by several investigators. *Van Der Kaaden* (1959) and *Schuiling* (1959) interpreted its formation by analogy to the Menderes Massif. According to them, the core consists of plagioclase-biotite gneiss which originated from old sediments after anatexis. In addition, there are olivine schist, amphibolite, and marble. The north-trending foliation in the core is unconformably cut off by the northeast-trending schistosity of the mantle rocks, which start out with thick greenschists and end with phyllite und marble layers. Some granodiorite plutons can be found around the gneissic core, but at least one of them is young (Eybek granodiorite, biotite, K/Ar, 35.9 ∓ 2 m. y., *Bürküt,* 1966: 125). *A. Gümüs* (1964) and *Aslaner* (1965) have investigated in detail the regional and contact-metamorphic as well as the eruptive rocks on the eastern slope of the massif.

Bingöl (1968), on the other hand, regards the massif as an anticlinorium consisting of several folds. The core of the central ridge is composed of dunite which, in stratigraphic sequence, is succeeded by amphibolite, metabasalt, gneiss, and marble; all of them are predominantly in amphibolite facies. From the east a block of greenschist and phyllite has been thrusted on the anticlinorium.

From a number of radiometric dates *Bingöl* (1971) developed the following geochronology for the massif:

25 m. y. ago – retrograde metamorphism
233 m. y. ago – tectogenesis and metamorphism up to anatexis, intrusion of granodiorite
304 m. y. ago – eruption of metabasalt.

According to those dates the formation of the ultrabasic and basic eruptive sequence had come to an end by the Middle Carboniferous. The tectogenetic and metamorphic main phase is regarded as Late Variscan. The dates are compatible with the geological evidence. The unmetamorphosed cover beds of the Kaz Dağ start on the western slope with Upper Carboniferous rocks and on the eastern slope with Triassic rocks (*Brinkmann,* 1971c: 63; *A. Gümüs,* 1964: 62; *Aslaner,* 1965: 39).

Ulu Dağ Massif

According to *Van Der Kaaden* (1959), this mountain range is petrographically very similar to the Kaz Dağ. Its core, too, consists of anatectic paragneiss, ranging from mesozonal to katazonal, with amphibolite lenses. No marbles can be found in the Ulu Dağ gneiss. They are present, however, in the epimetamorphic sequence of the mantle accompanied by mica schist, quartz phyllite, graphite phyllite, and quartzite schist.

The foliation of the gneissic rocks runs mainly north. Later, gneiss and mantle rocks both became folded and exhibit west-southwest-trending folds which are verging to the south (*Ketin,* 1947b; *Lisenbee,* 1971: 361, 1972: 69). The youngest rocks of the basement are discordant granite plutons which predominantly have spread out in the main anticline.

The regional metamorphism and folding of the massif has taken place before the intrusion of the granite. Its radiometric age is:

total rock	Rb/Sr	245 ∓ 37 m. y.	(*Öztunalı,* 1973: 1)
total rock	K/Ar	269 ∓ 39 m. y.	
orthoclase	Rb/Sr	235 ∓ 35 m. y.	
biotite	Rb/Sr	30 ∓ 3 m. y.	
biotite	K/Ar	24 m. y.	(*Bürküt,* 1966: 53)

From a stratigraphical point of view, the crystalline complex is of pre-Upper Carboniferous age (*Brinkmann,* 1971c : 63). Therefore, it appears that the Ulu Dağ and the Kaz Dağ Massifs became covered with unmetamorphosed sediments even before their cores had been completely cooled off.

Alanya-Anamur Massif

The massif represents only the northeastern part of a crystalline dome underlying the Levantine Sea (*Blumenthal,* 1951; *De Peyronnet,* 1971). Along the coast higher-grade metamorphic rocks, garnet-bearing mica schist and hornblende schist crop out. Farther inland they are succeeded by chlorite schists with quartzite beds, sericite phyllite, and marble. All rocks belong to the greenschist facies. As in the Bitlis Massif, this basement is covered – unconformably and with metamorphic hiatus – by thick crystallized limestones which extend from the Permian to the Triassic. Their degree of recrystallization decreases towards the northeast.

This massif, too, has undergone two metamorphic phases. The older phase seems to have taken place in the Paleozoic, but its precise timing is controversial. *Striebel* (1965: 17) reports Upper Carboniferous conodonts from the uppermost levels of the crystalline base; whereas, according to *Işgüden* (1971: 31), even unmetamorphosed Upper Devonian sediments are overlapping the base at Anamur. The age of the younger metamorphic phase is uncertain too. *De Peyronnet* (1971: 95) does not commit himself if in addition to Permian also Triassic rocks were affected.

Summary

The classification of the crystalline schists of Turkey with respect to age and origin is still in its infancy, due to the limited number of petrological investigations and radiometric dates. A clear separation of the metamorphites into Cryptozoic and Phanerozoic units is not yet possible. So far, the gneiss and amphibolite of the Bitlis Massif and the gneiss of the Istranca Massif are the only rocks in Turkey whose age can tentatively be assigned as Precambrian.

Parallels between the individual massifs should be drawn with precaution. Therefore, the question cannot be answered whether the gneissic rocks, with north-trending foliation, which occur in the cores of several western Anatolian massifs belong to a Lower Paleozoic or Proterozoic tectogene (*Schuiling,* 1962: 79). Likewise, the question of whether the thick greenschist sequences in the Kaz Dağ and Ulu Dağ are equivalent to the Bulgarian diabase-phyllite formation remains unanswered. On the other hand, it seems justified to reconstruct a south-verging fold arc, which is open to the south, based on the structure of the younger metamorphites of the Kaz Dağ, Ulu Dağ, and the northern Kırşehir Massif. The time of formation of this arc is uncertain.

Many more radiometric dates are available for the Phanerozoic than for the Cryptozoic epoch. In fact, in many cases it is not clear yet whether the data are applicable to the age of origin, of regional metamorphism, or of a later reheating. Geological observations indicate, however, that metamorphic phases and intrusions of Variscan and Alpidic origin have definitely taken place (p. 38, 95). Accordingly, at least three periods of regional metamorphism can be distinguished in Turkey: an Early Paleozoic or pre-Paleozoic, a Variscan, and an Alpidic period.

Chapter 4 Infracambrian and Cambrian

Fossil-bearing Cambrian rocks have first been found in southeastern Anatolia by *Tolun* and *Ternek* in 1952. It is likely that those rocks are rather widespread in the south of Anatolia, however, only a few profiles have been thoroughly investigated. The same holds true for the fauna. Only trilobites have been described (*Dean* and *Krummenacher,* 1961; *Dean,* 1972). Summaries of the system have been prepared by *H. Flügel* (1964, 1971), *Ketin* (1966b), *Wolfart* (1967b), and *Haude* (1969).

The Infracambrian and Cambrian in Southwest and South Anatolia (Fig. 4)

The base of the Cambrian is exposed only west of Beyşehir Gölü as slightly metamorphosed phyllite (*Dumont,* 1972). It is conformably overlain by quartzite and sandy shale, which are also present in the Sultan Dağ (*Haude,* 1972: 413). Only traces of fossils can be found. Next comes a sequence of carbonate rocks (*Dean* and *Monod,* 1970: 416; *Haude,* 1972: 414) which is composed of gray and black crystalline limestone and dolomite in its lower part, and of red nodular limestone in the upper part. The dark rocks carry almost exclusively stromatolites. The red limestone which is turning into marl at the hanging wall, contains a Middle Cambrian trilobite fauna. At the Beyşehir Gölü, in the Sultan Dağ, and south of the Sultan Dağ near Hadım, it is succeeded by uniform, grayish-green flyschoid shale with sandstone beds. The lower part of the strata contains Middle Cambrian, the middle one Upper Cambrian, and the highest one Ordovician fossils (*Özgül* et al., 1972: 15; *Özgül* and *Gedik,* 1973).

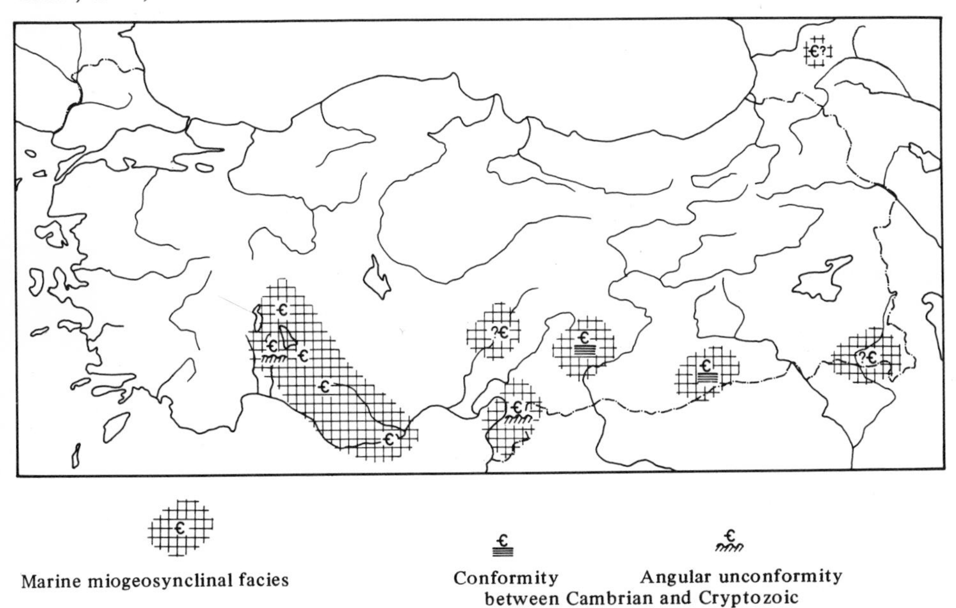

Fig. 4 Distribution, facies and paleogeography of the Cambrian in Turkey

Cambrian beds probably occur in the eastern Taurus because lowermost Ordovician strata have been found near Ovacık west of Silifke (*Yalcınlar,* 1973: 104) and in the Antitaurus (*Özgül* et al., 1972, 1973).

The Infracambrian and Cambrian in Southeast Anatolia (Fig. 4)

In the Amanos Dağ and the area of Adıyaman and Mardin the lowest Cambrian and its base is different than in southwestern Anatolia. In the Amanos Dağ, the oldest rocks are thin beds of fine-grained quartzitic sandstone and slightly phyllitic schist which probably belong to the latest Proterozoic. They are overlain in angular unconformity (*Dean* and *Krummenacher,* 1961: 73; *Atan,* 1969: 30; unpub. data) or disconformity (*Ketin,* 1966b: 84; *Janetzko,* 1972: 5) by slightly reddish, coarse-grained, partly conglomeratic quartzite and arkose sandstone, which are believed to be of Lower Cambrian age. Near Adıyaman and Mardin there are outcrops of 3000 m thick beds of red cross-bedded continental sandstones with intercalations of andesite and basalt. *Ketin* (1966b: 78) assigns this sequence partly to the Proterozoic and partly to the Lower Cambrian with the boundary at a place where the volcanic intercalations are decreasing markedly. The higher part of the system in southeastern Anatolia is formed along the same lines as in southwestern Anatolia. The Middle Cambrian series is calcareous-dolomitic and rich in fossils, and the fine-clastic Upper Cambrian series contains worm tubes and brachiopod coquina.

In the core of the Ricgar anticline close to the Iraqi border, *Altınlı* (1966: 54) found a huge series of dark bedded limestones. They underly a sequence of possibly Ordovician age and may belong to the Cambrian. *S. Türkünal* (1953: 9), however, mentions only Ordovician (?) quartzite and phyllite for that area (p. 17).

Summary

The base of the Cambrian in Turkey consists of unmetamorphosed to slightly metamorphosed rocks. Similar rocks occur in Iran and are considered Precambrian/Cambrian transition beds (*Stöcklin* et al., 1964).

The oldest Cambrian sediments probably are found in Southeast Anatolia. There the red beds might represent the evaporite-free marginal facies of the Iranian evaporation basin where the Hormuz salt series became precipitated. According to *Stöcklin* (1968b: 164), it belongs to the Lowest Cambrian or even the Upper Proterozoic. The basin may have been the embryonic stage of the Cambrian Tethyan geosyncline and at the same time the starting point of the Cambrian transgression in Asia Minor (*Wolfart,* 1967b: 170). From the upper Lower Cambrian the entire area became covered by a shallow sea in which the same sediment sequence was deposited from Spain to Iran.

So far, the existence of the Cambrian Tethys can only be proven for the southern part of Anatolia. There, a miogeosyncline existed until Ordovician time with 1000 to 2000 m thick sediments which include a few lava beds. Marine Cambrian should be expected in northern Anatolia also, for east and west of Turkey, i. e. in the High Caucasus (*Tchernisheva,* 1968) and in the Balkan Mountains (*Kalenić,* 1967) Cambrian beds are present. It is conceivable that the lower part of the red beds of the northwest Anatolian "Ordovician" is of Cambrian age.

The fauna of the Mediterranean Cambrian is as uniform as its lithology. The main fossil horizon of the Turkish profiles is the Middle Cambrian. Its trilobite fauna is closely related to that of Morocco, southwestern and Central Europe. It shows a truly atlantic character, although the boundary of the Redlichia province is not far away, as shown by the evidence found in Israel and Iran (*Kushan,* 1973; *Wolfart* and *Kürsten,* 1974).

Cambrian and Infracambrian

	W. Taurus, Sultan D. (Dean + Monod 1970, Haude 1969, 1972, Dumont 1972)	Amanos Dağ (Dean + Krummenacher 1961, Atan 1969)	Amanos Dağ (Janetzko 1972)	Derik near Mardin (Ketin 1966)
Overlying	> 1000 m	Ordovician		
Upper Cambrian	Seydişehir – Formation	300-700 m White quartzite	500 m Tandır Formation	1000 m Quartzite
Middle Cambrian	Greenish gray shale with sandstone beds; 10 m Buff marl	Mekersin (Tiyek) Formation: Arenaceous shale + sandstone; 10 m Nodular limestone	Gray quartzite + siltstone + shale	Sosink Formation: Marl + shale + sandstone; 30 m Red nodular limestone
	100 m Çaltepe Form.: 10-40 m Red nodular lms.; 50 - 100 m Gray lms.+dol.	150 m Karayüce Form. Gray dol. limestone	Kayabaşı Formation: 10 m Red + yellow marl; 150 m Gray cryst. lms.	250 m Dolomite Form.
Lower Cambrian	> 100 m Ardıçlıtepe Form. Dark gray shale with quartzites	250 m Eğrek Form. Gray + reddish congl. quartzite	400 m Kuşçu Formation Coarse conglomeratic quartzite	700 m Sadan Formation Red sandstone + siltstone + conglomerate
Infracambrian	> 2000 m Sarıçiçek Form. Greenish gray phyllite	> 1000 m Quartzite + phyllitic shale	Tesbi Formation Fine grained quartzite + shale	> 2000 m Telbismi Formation Red sandstone + siltstone + andesite + basalt

In the stratigraphic tables continental deposits are stippled.

Chapter 5 Ordovician

Since *Tchihatcheff's* travels, it is known that multicolored sandstones are quite abundant east of the Bosporus, on the Kocaeli Peninsula. *Kessler* (1909) and *Endriss* (1910) proved that they have been formed prior to the Silurian. Their stratigraphic position was uncertain till the fossil findings of *Arıc-Sayar* (1955). *Fuchs* (1902) first reported Ordovician in the south of Turkey. In addition to trilobites (*Dean,* 1967, 1971, 1973), only fossil markings (*Fuchs,* 1902; *Broili,* 1911) and Conularia (*Sayar,* 1969) were investigated closer. For recent summaries see *Paeckelmann* (1938: 180), *H. Flügel* (1964, 1971), *Wolfart* (1967b), and *Haude* (1969).

The Ordovician of Kocaeli (Fig. 5)

The oldest member of the Paleozoic is represented in outcrops of multicolored, feldspar-bearing clastic rocks. Their base is unknown. The pebbles, ranging in diameter from cm to dm, are mostly composed of quartzite, milky quartz, jasper, and more rarely of phyllite, mica schist, gneiss and granite (*Udluft,* 1939: 6; *Okay,* 1947: 271; *Altınlı,* 1951b: 192; *Haas,* 1968a: 186). The red formation is conformably succeeded by the light, silicic Ayazma strata. Cruziana trails and brachiopods are the only fossils found in the quartzites. The siliceous shale contains a chamosite-oolite bed in which Conularia and asaphids of the Middle Ordovician epoch are present (*Yalcınlar*, 1956; *Baykal* and *Kaya*, 1965: 6; *Sayar*, 1969). If the position assigned to the chamosite bed is correct, then – contrary to *Haas* (1968a: 187) – at least all of the Ayazma strata should be placed in the Ordovician system.

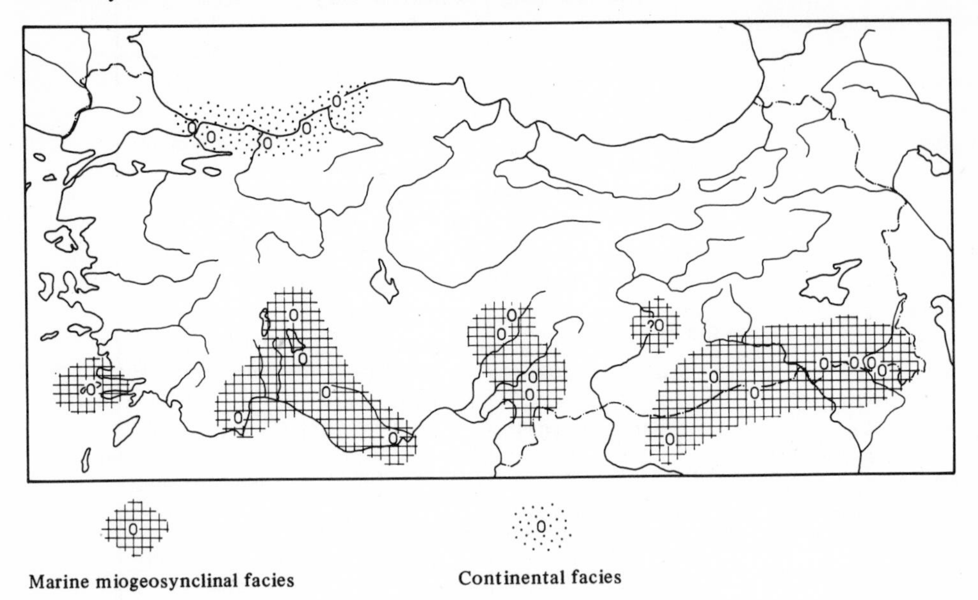

Fig. 5 Distribution, facies and paleogeography of the Ordovician in Turkey

At the beginning of the Ordovician period, and possibly already in the Cambrian (p. 14), a continental basin existed on Kocaeli. It collected fluviatile, fanglomeratic debris of crystalline source. During the Middle Ordovician, the sea intruded into the basin, but the facies remained littoral-neritic. The Ayazma strata contain a number of quartzite lenses which may have been ancient beach ridges. The origin of the clastic material and the direction of transgression is not yet clear. *Haas* (1968a: 192) assumes an area of denudation north of Kocaeli.

The Ordovician in the Remainder of Northwest Anatolia (Fig. 5)

Nonfossiliferous sequences similar to the Ordovician lithologic facies of Kocaeli, are exposed underlying Devonian beds in several anticlines east of the lower course of the Sakarya. Grain size and feldspar content is gradually decreasing towards the east. The colors become less intense, and limestone intercalations appear. In the Çam Dağ near Adapazarı (unpub. data) and in the Bolu Dağ, there are several 1000 m thick strata of gray and reddish, partly conglomeratic sandstones. They probably rest on a phyllitic basement in the Çam Dağ, and on gneiss in the Bolu Dağ with a thick intercalated bed of diabase (*Niehoff*, 1960). Near Ereğli, the sandstone of Hamzafakılı belongs to this sequence (*Tokay*, 1952: 42). Finally, the multicolored, sandy shale and light quartzite with brachiopod coquina of Inkum near Bartın (*Charles*, 1933: 75; *Tokay*, 1954/55: 47) should be the equivalent to the Bithynian series.

The Ordovician in West Anatolia

The occurence of Ordovician on the Karaburun Peninsula, as suggested by *Höll* (1966: 30), is not adequately documented.

The Ordovician in Southwest and South Anatolia (Fig. 5)

In the western part of the Taurus and its northern hinterland as well as at Kos Island (*Desio,* 1931: 143), the Ordovician is represented by uniform sandy shales with sandstone beds. They contain in the Sultan Dağ (*Haude,* 1972: 415) and south of these mountains near Hadım (*Özgül* and *Gedik,* 1973) a Tremadocian fauna, and near Seydişehir (*Dean* and *Monod,* 1970) an Arenigian fauna. In the basement of the western Taurus, Ordovician sediments must be widespread, for they reappear in a thrust slice south of Antalya (*Brunn* et al., 1972a: 238).

According to *Yalcinlar* (1973: 104), similar sandy-clayey marine sediments near Ovacik west of Silifke contain a few, only tentatively identified fossils of the Lower Ordovician.

Formerly, there were only two occurrences of Phycodus circinatus known in the Antitaurus (*Fuchs,* 1902; *Broili,* 1911: 2). Recently, *Özgül* et al. (1972, 1973: 88) discovered the corresponding strata in the form of 1000 m dark shales and quartzites with fossils from the Tremadocian and Arenigian stage.

The Dictyonema which *Yalcınlar* (1963: 17) mentions from the crinoid-bearing metamorphic schist of the Baba Dağ near Denizli, may well be pseudo-fossils. .

The Ordovician in Southeast Anatolia (Fig. 5)

Two facies can be distinguished here. The Ordovician of Bedinan (*Dean,* 1967), which is adjacent to the Cambrian at Mardin (p. 14), shows the same fine sandy-clayey lithology as in Southwest Anatolia, and as the strata found in the boreholes of Abba and Kamichlie in northern Syria (*H. Flügel,* 1964: 12). The second facies is characterized by coarser-clastic sediments, especially in the Lower Ordovician. In the Amanos Dağ, Cruziana quartzite is prevalent in the Arenigian stage, and dark shales with quartzite and sandstone beds in the Caradocian stage (*El Ishmawi,* 1972: 37). The Ordovician in the extreme southeast of Turkey has an even stronger littoral character. In the cores of the Büyük Zap anticlinorium east of Harbol and in the Ricgar Mountains south of Hakkâri, thick layers of gray and red quartzitic sandstone with Cruziana (?), overlain by Devonian or Permo-Carboniferous sediments, are exposed (*S. Türkünal,* 1953: 9; *Altınlı,* 1954b: 40). They contain tuff layers (?) and correspond, according to *G. C. Schmidt* (1964: 106), to the Iraqi Khabour formation in lithology and age.

Summary

Although occurrences of Ordovician rocks are still sparse in Turkey, a symmetrical arrangement of the facies is evident. Multicolored, coarser-clastic rocks of partly continental and partly littoral origin are abundant in the northwestern as well as in the southeastern part of Turkey. They are divided by a belt of uniform, fine-sandy marine sediments extending from Kos in the west to Mardin in the east. The paleogeographic framework thus created is that of an east-west oriented strait with shorelines on both sides that covered South Anatolia. To the north, it was bordered by an area of denudation, the *Pontian land* (*Frech,* 1899), situated in the present Black Sea. Its southern shore along the Afro-Arabian craton was accompanied by a wide and sandy zone which extended from Syria across Israel, Jordania, to Iraq.

The thickness of the Ordovician strata corresponds to that of other geosynclines; they exceed 1000 m in the marine as well as in the continental facies. Volcanic activity was almost non-existent.

The relationship of the marine fauna points, according to *Dean* (1971: 20, 1973: 337), primarily to Bohemia and North Europe, and less pronounced to Burma and China. It confirms that at that time Anatolia was part of the Tethys.

Ordovician

	Kocaeli (Haas 1968 a) — Bithynian series	W. Taurus, Sultan D. (Dean 1971, Haude 1972)	Antitaurus (Özgül et al. 1973)	Amanos Dağ (El Ishmawi 1972, Lahner 1972)	Derik - Mardin (Dean 1967)	Hakkâri (Altınlı 1954 a, b)
Overlying	Akviran series	Triassic	Silurian	Silurian	Cretaceous	Devonian
Ashgillian	200 m Yayalar Formation, Graywacke + sandstone		150 m Halit Yaylası Formation, Shale + sandstone + conglomerate	200m Ayran (Kıslaç) Formation, Gray arenaceous shale with sandstone beds		
Caradocian	250 m Siliceous shale +				500m Bedinan Formation, Shale + ss.	1000 m Giri Formation, Gray and variegated quartzite
Llandeilian	Ayazma Form. iron oolite; White + reddish quartzite		1200 m Armutludere Formation, Arenaceous shale + quartzite		?	
Llanvirnian	Quartz conglomerate			500 m Bahçe Formation, White quartzite + greenish sandstone		
Arenigian	> 2000 m Kurtköy Formation, Gray + variegated graywacke + siltstone + conglomerate	30 m Sobova Form., Limestone + var. sh				
Tremadocian		> 1000 m Seydişehir Formation, Shale + graywacke				
Underlying	?	Middle Cambrian	? Cambrian	Cambrian	Cambrian	?

Chapter 6 Silurian

When in 1864, *De Verneuil* investigated fossils from the Istanbul region, he noticed not only Devonian but some Silurian species too. In the south of Turkey, the Silurian system was first mentioned by *H. Flügel* and *Yalcınlar* in 1955. Reports on the Silurian paleontology of Kocaeli have been published by *Paeckelmann* (1938) and *Haas* (1968). The graptolite fauna has been treated by *H. Flügel* (1955a). Summaries on the Silurian of Turkey have been given by *Paeckelmann* (1938: 183), *H. Flügel* (1964, 1971), *Wolfart* (1967b), *Haude* (1969), and *Berry* and *Boucot* (1972).

The Silurian of Kocaeli (Fig. 6)

The strata have recently been investigated by *Haas* (1968a: 187). Graywacke and chamositic shale are representative for the transition zone extending from the Ordovician to the Silurian. At the end of the Llandoverian stage, clastic sedimentation diminished and carbonates were deposited instead. At the beginning, marls and fine-sandy limestones were formed. Starting with the upper Wenlockian stage, organogenic limestones, crinoidal and Pentamerus coquina, as well as limestones with stromatoporids, tabulate and rugose corals became quite abundant. However, clastic material still reached the sea in such quantities that no real reefs could form. At the end of the Silurian, marly limestones and finally black banded bituminous limestones poor in fossils began to appear.

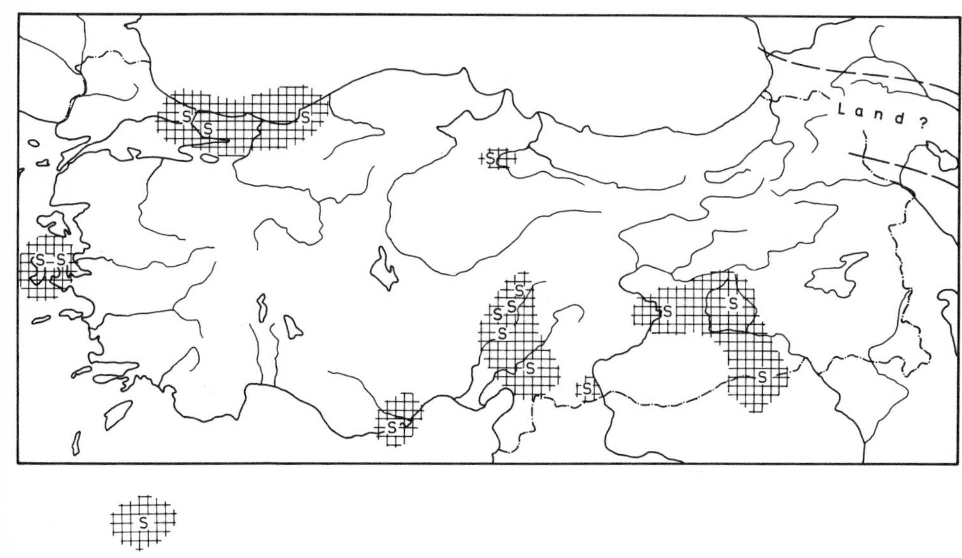

Marine miogeosynclinal facies

Fig. 6 Distribution, facies and paleogeography of the Silurian in Turkey

This facies change reflects a progressive transgression as indicated by the reduction of terrigenous material. Until the Ludlovian stage, there was an epicontinental sea, abundant with marine life. At the end of the Silurian, the sea deepened considerably.

During all of the Silurian period, the northwestern part of Kocaeli was closest to the shore, according to *Paeckelmann* (1938: 150) and *Haas* (1968a: 211).

The Silurian in the Remainder of Northwest Anatolia (Fig. 6)

In the Çam Daǧ and near Bartın, the Silurian system is missing. It is preserved only at two localities. At Eregli the Ordovician is overlain by thinly laminated sandy marl with brachiopods and graptolites (*Egemen,* 1947; *Tokay,* 1952: 42). Southeast of Amasya, 2000 m-thick layers of graywacke and shales are resting directly on crystalline rocks; and limestone lenses contain cephalopods of Middle Silurian age (*Alp,* 1972: 25).

The Halysites locality of Kalecik north of Ankara (*Lebling,* 1925: 106) has never been re-confirmed.

The Silurian of West Anatolia (Fig. 6)

In the northern part of the Karaburun Peninsula and on Chios Island, outcrops of graywacke and shales with chert and limestone lenses are abundant. Its oldest parts in both areas are of Silurian age. On Karaburun, the limestone of Kalecik contains an upper Ludlovian-Gedinnian fauna which is equivalent to the limestone of Agrelopos on Chios (*Höll,* 1966:32; *Besenecker* et al., 1968: 128; *Lehnert-Thiel,* 1969: 49).

The Silurian of Southwest and South Anatolia (Fig. 6)

In contrast to the wide range of Ordovician sediments, Silurian localities are quite rare. In the Sultan Dağ, 1000 m of dark, fine-sandy shales seem to fill the time gap from the Tremadocian to the Upper Devonian. They do not contain fossils other than (?) graptolites (*Yalcınlar,* 1959; *Haude,* 1972: 415).

Rich graptolite localities are situated, however, in the eastern Taurus (Ovacik west of Silifke) and the Antitaurus (Feke and south of Pınarbaşı). The Llandoverian stage is represented by typical graptolite shale. It is underlain by sandy shales with conglomeratic beds, and is succeeded by fossil-rich Orthoceras limestone and nodular limestone (*H. Flügel,* 1955a; *Yalcınlar,* 1973: 108; *Özgül* et al., 1973: 89), which according to new investigations (*Buggisch,* 1973), represent the transition from Ludlovian to Gedinnian.

The Silurian in Southeast Anatolia (Fig. 6)

Here again, the Silurian system is only sporadically present. Outcrops of fine-sandy shales with monograptids are known near Çüngüş at the Frat (*Tolun,* 1962: 221; *Altınlı,* 1966: 55). Similar sandy-clayey strata, known as Daday or Hanof formation, were found in boreholes in the Hazro, Diyarbakır, and Gaziantep area (*Berry* and *Boucot,* 1972: 51).

In the Silurian of the Amanos Dağ, multicolored conglomerates, sandstones and shales are prevalent; their lower part contains a few brachiopods of the Llandoverian stage (*El Ishmawi,* 1972: 40; *Lahner,* 1972: 70).

It is possible that the gray and reddish, micaceous fine sandstone beds, which are overlying the multicolored Ordovician rocks in the Büyük Zap anticlinorium south of Hakkâri, belong to the Silurian system (*Altınlı,* 1966: 55).

Summary

The distribution of the Ordovician and the Silurian in Turkey are similar insofar as both systems are almost missing in Middle Anatolia. The question has to remain open of whether this phenomenon was caused by a gap in sedimentation, by later erosion, or by regional Caledonian metamorphism (p. 22).

The facial pattern of the Silurian reflects to a certain extent that of the Ordovician. Where littoral or continental multicolored sediments are abundant in the Ordovician system, neritic conditions were dominant in the Silurian system. Where sediments of a shallow sea were deposited in the Ordovician, graptolite shales were formed, at least during the earlier Silurian epoch. The major structural framework in Turkey, therefore, remained almost unchanged from the Ordovician to the Silurian, the whole area had only subsided during the Silurian. It was not until the end of the Silurian period that a change took place. In the latest Ludlovian, bathyal sediments are found on Kocaeli, whereas coquina limestone is present in the Antitaurus.

The thickness of the Silurian beds is about 500 m; i. e. less than the other Paleozoic systems. Flyschoid sediments as well as volcanic intercalations are missing. The geosynclinal subsidence was relatively quiet.

The paleobiogeographic relations remained the same as before. The Silurian as well as the Ordovician marine fauna of Turkey is closely related to that of Europe (*Berry* and *Boucot*, 1972: 45).

Silurian

	Kocaeli Haas 1968 a		Antitaurus Özgül et al. 1973 Buggisch 1973	Amanos Dağ El Ishmawi 1972 Lahner 1972
Overlying	Marmara series		↑	Devonian
Ludlowian	200 m Kireçhane Form. Black bedded limestone	Akviran series	150 m Yukarıyayla Formation Orthoceras limestone + gray marl	500 m 250 m White + red sandstone + quartzite
	70 m Pelitli Formation Nodular limestone			
	40 m Çakıllıdere Formation Brachiopod coquina			
Wenlockian	40 m Cumaköy Formation Pink crinoid limestone			Dedeler (Çerleme) 250 m Red siltstone + shale
	50 m Bağlarbaşı Formation Arenaceous limestone			
	50 m Tavşantepe Formation Marl			
Llandoverian	200 m Yayalar Formation Graywacke + shale	Bithynian series	100 m Pusçutepe Formation Black graptolite shale + lydite	Formation 50 m Variegated coarse quartzite
Underlying	Ayazma Formation		Ordovician	Ordovician

Chapter 7 Caledonian Tectogenesis

The question of Caledonian tectogenic movements in Turkey has first been raised by *Ketin* (1953, 1959b).

The Caledonian Tectogenesis in Northwest Anatolia

No full accordance has been reached for Kocaeli. On the one hand, *Haas* (1968a: 192) is pointing out that from the Ordovician to the Devonian all profiles studied by him are conformable, except for a reduced thickness or a break in the Gedinnian or Siegenian stage. On the other hand, the Devonian system is definitely overlapping the Kurtköy strata with angular unconformity near Kartal (*Abdüsselâmoğlu,* 1963; unpub. data). At the summit of the Büyük Çamlıca, too, deposits of the Emsian stage are found next to Lower Silurian sediments. Since, however, the contact is not exposed, the relations are interpreted in different ways (*Paeckelmann,* 1938: 88; *Altınlı,* 1954c: 219; *Ketin,* 1959a: 16). The contradictions can be resolved by assuming that the Caledonian folding was gentle. On the anticlines, the Ordovician became exposed by erosion, and in the synclines, the sedimentation continued without marked interruption. The strike of the folds is uncertain; according to *Ketin* and *Abdüsselâmoğlu* (1963: 6), it is to the east. As to timing, only one of the late Caledonian phases is possible, most likely the interval between the Gedinnian and Emsian stages which was almost devoid of sedimentation.

For the remainder of Northwest Anatolia, there is no question concerning the validity of Caledonian movements. In the Çam Dağ near Adapazarı exists an angular unconformity of up to 50° between the Ordovician and the Lower Devonian system (unpub. data). Near Ereğli, the Emsian stage rests partly on Ordovician and partly on Lower Silurian strata (*Tokay,* 1952: 43). In the Bolu Dağ, the Devonian is overlapping Ordovician and crystalline rocks (*Batum,* 1968: 14).

The Caledonian Tectogenesis in South and Southeast Anatolia

Caledonian unconformities have been reported at several places in the eastern Taurus and Antitaurus. They are supposed to occur partly between the Lower Ordovician and Silurian and partly between the Silurian and Upper Devonian (*Özgül* et al., 1972: 10; *Yalcınlar,* 1973: 109). The rest of the Early Paleozoic sequence in the southern and southeastern part of Anatolia is conformable. However, the Silurian is frequently missing, which is probably due to uplift and erosion that occurred simultaneously with the Caledonian orogeny. Thus, south of Hakkâri, Lower Devonian may be deposited on partly eroded Silurian strata (*Altınlı,* 1966: 55). Near Harbol and at the Ricgar Dağ it rests directly on Ordovician rocks (*S. Türkünal,* 1953: 9; *Altınlı,* 1954b: 41). In the Amanos Dağ as well, the profiles are partly stratigraphically continuous and partly incomplete (*Dean* and *Krummenacher,* 1961: 73; *El Ishmawi,* 1972: 58; *Lahner,* 1972: 92).

Summary

In conclusion, a Caledonian tectogenesis has occurred in Northwest Anatolia, but it was a weak one. In the south of Turkey, it was primarily developed in the form of epirogenic movements. Angular unconformities reported are obviously only of local importance. Indications of regional metamorphism during the Caledonian epoch are not known in Turkey. Thus, there is no reason to explain the conspicuous lack of the Cambrian and Ordovician systems in Middle Anatolia with a Caledonian recrystallization (p. 20).

Chapter 8 Devonian

In 1848, *De Tchihatcheff* found the first Devonian fossils near Istanbul as well as in South Anatolia. The paleontological literature about the area surrounding Istanbul is presented by *Paeckelmann* (1925, 1938), *Haas* (1968), and *Kaya* (1973). For the remainder of Turkey see *Penecke* (1903), *Broili* (1911), *Frech* (1916), *Heritsch* (1928), *Heritsch* and *Von Gaertner* (1929), *Charles* (1933), *Delépine* (1933), *Le Maître* (1930, 1933), *Ünsalaner* (1951), *H. Flügel* (1955b), and *E.* and *H. Flügel* (1961). Geological summaries were prepared by *Paeckelmann* (1938: 184), *H. Flügel* (1964, 1971), and *Wolfart* (1967b).

The Devonian at Istanbul and on Kocaeli (Fig. 7)

This area contains the best studied and most complete Devonian profiles of Turkey. *Paeckelmann* (1925a: 1) wrote its history of investigation, and recently *Haas* (1968a) and *Kaya* (1973) reclassified the strata.

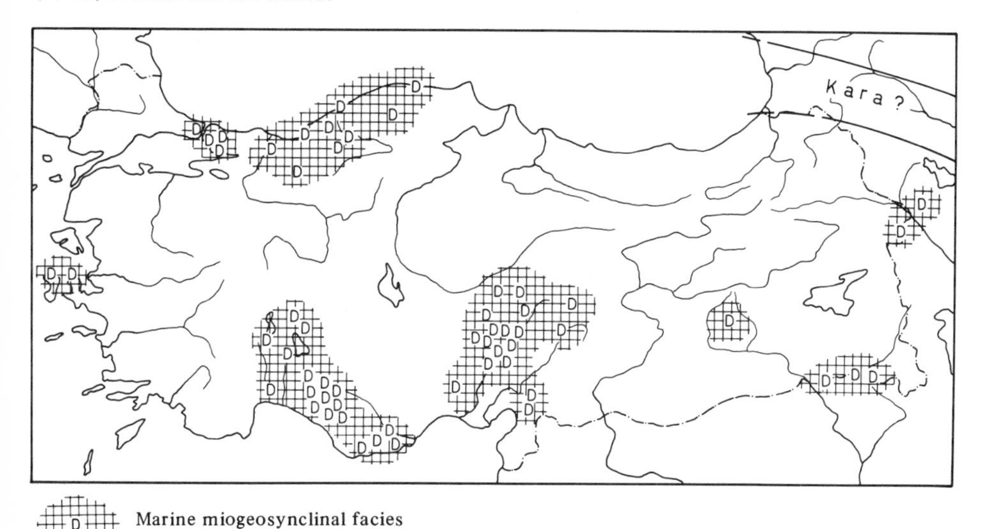

Fig. 7 Distribution, facies and paleogeography of the Devonian in Turkey (Kara = Land)

Phacoidal limestones extend without interruption from the uppermost Ludlovian to the Siegenian stage. The upper Lower Devonian rocks, however, are mostly arenaceous. At the base, they consist of calcareous graywacke shale, followed by fossil-rich calcareous sandstone and calcareous graywacke, and finally fine-sandy marly limestone. The Middle and Upper Devonian is composed of marly shales and nodular limestones with poor fossil content. By uptake of siliceous nodules, they develop into the chert beds of the Lower Carboniferous.

During the Lower Devonian period, Kocaeli was almost entirely covered by a shallow sea. *Haas* (1968a: 224) shows how the sediments and fauna of the Rhenish and Hercynian facies succeed each other in time and space. He concludes, as did *Paeckelmann* (1938: 153), that during this period the northeastern part of Kocaeli was closer to shore than the southwestern part. Starting with the Middle Devonian and extending into the Carboniferous period, an expansion and deepening of the sea took place which blurred the continental influence from the north.

The Devonian in the Remainder of Northwest Anatolia (Fig. 7)

The localities can be divided into three zones. The deposits of Ereğli (*Heritsch* and *Von Gaertner,* 1929; *Tokay,* 1952: 43), Bartın (*Charles,* 1933; *Delépine,* 1933; *Tokay,* 1954/55: 47, 1962: 3), and Inebolu (*Heritsch* and *Von Gaertner,* 1929) are close to the shore of the Black Sea. Near Ereğli and Bartın, sandy-shaly Lower Devonian sediments with basal conglomerates are overlapping Ordovician and Silurian deposits, respectively. Younger Devonian strata are missing at Ereğli, but they are present south of Zonguldak (*Altınlı,* 1951a: 159). Near Bartın, the Middle and Upper Devonian consists of thick sequences of bedded to reef-like limestone and dolomite which continue into the Lower Carboniferous system.

A little farther inland, Devonian outcrops occur near Adapazarı and Bolu. Here, too, the Devonian is closely associated with the Ordovician. Near the boundary between the Lower and Middle Devonian system, the profile of the Çam Dağ contains a 25 m-thick, low-grade iron oolite bed (*Van Wijkerslooth* and *Kleinsorge,* 1940; unpub. data). In the Bolu Dağ and Arkot Dağ, the Paleozoic is quite abundant (*Blumenthal,* 1948: 156, 164), however, only Lower and Middle Devonian deposits have been reported (*Heritsch* and *Von Gaertner,* 1929: 199; *Batum,* 1968: 10).

Even more distant from the Black Sea the deposits of Mudurnu (*Abdüsselâmoğlu,* 1959: 27), Karabük (*Nowack,* 1928: 309), and Daday (*Grancy,* 1939: 86) are situated. In this zone, the Devonian is overlapping crystalline schists. A small Lower Devonian fauna has been found near Mudurnu in the sandy shales which overlie the basal arkosic conglomerate.

In Northwest Anatolia, the Devonian transgression thus emanated from Kocaeli which is the only area that shows a complete Silurian-Devonian section. From there on it advanced toward the east parallel to the shore of the Black Sea. It followed along an older geosynclinal basin which contained Ordovician, and occasionally Silurian sediments and which was accompanied by a crystalline uplift in the south.

The Devonian in West Anatolia (Fig. 7)

Parts of the graywacke and shales at Karaburun, as well as at Chios, are probably of Devonian age.

At Soma, limestone pebbles, taken from the Orhanlar graywacke (p. 29), contain Upper Devonian conodonts (*Brinkmann* et al., 1970: 10).

The crinoid-bearing phyllites of the Baba Dağ near Denizli (p. 8) may be of Devonian origin.

The Devonian in Middle, Southwest, and South Anatolia (Fig. 7)

Haude (1972: 415) considers the Early Paleozoic sequence of the Sultan Dağ a complete one, even though no paleontological evidence has been found for the Silurian or the older Devonian. Only the Upper Devonian is documented, and in spite of its partly conglomeratic facies, it is assumed to fit conformably into the profile.

In the basement of the western Taurus, the Devonian extends over large areas, as shown by several outcrops near Konya (*Wiesner,* 1968: 180), between Eğridir and Beyşehir Gölü (*Ünsalaner,* 1941; *Altınlı,* 1944: 228; *Brunn* et al., 1972a: 521), and the eastern part of the mountain range at Seydişehir, Akseki, and Alanya (*Blumenthal,* 1944b: 106, 1949: 18, 1951: 39). At Konya Silurian-Devonian boundary strata have been reported, elsewhere Upper Devonian coral limestones and coquina limestones are exposed.

The Devonian is widespread in the eastern Taurus and, in particular, the Antitaurus. However, despite favorable exposures, no complete sections have been described (*Blumenthal,* 1941: 61, 1947b: 51, 1951: 39, 1952a: 42; *H. Flügel,* 1955b; *E.* and *H. Flügel,* 1961; *Özgül* et al., 1973). The Lower Devonian is rare, the Middle Devonian more frequent, and the Upper Devonian (mostly Adorf stage) is abundant. This disparity reflects, in part, the relative fossil content. It could have been caused, however, by an overlapping of the higher Devonian series. Indications for such a possibility exist. At Yahyalı south of Kayseri (unpub. data), at Bünyan east of Kayseri (*Baykal,* 1945: 136), and east of Anamur (*Isgüden,* 1971: 31), Lower to Upper Devonian beds are resting on crystalline rocks. Volcanites and tuffs are generally sparse in the Devonian. They are only mentioned from the Middle and Upper Devonian of the northwestern part of the Antitaurus (*Vaché,* 1964: 93; *Buggisch* et al., 1974).

A finding of Rhynchonella cuboides south of Ankara (*Leuchs,* 1943: 59) has not been reconfirmed. The "Devonian" found between Alanya and Gazipaşa is Permian in age, according to *Striebel* (1965: 20) and *Gedik* (pers. com.).

The Devonian in Southeast Anatolia (Fig. 7)

The Devonian of the Amanos Dağ shows like the lower Paleozoic mainly clastic formations. The strata markedly decrease in thickness from north to south, and the facies changes from neritic to littoral. The Lower Devonian series appears to be missing. The Upper Devonian is rich in fossils and lithologically closely associated with the Lower Carboniferous system (*El Ishmawi,* 1972: 41; *Lahner,* 1972: 72).

In the remainder of southeast Anatolia, dark shale, marl, and limestone are predominant; multicolored quartzite and shale are only found at the base. According to the few fossil localities, the marine development reaches from the upper Early Devonian to the lower Late Devonian epoch (*Chaput,* 1936: 238; *Mercier,* 1951; *S. Türkünal,* 1951: 36, 1953: 9, 1955: 52; *Altınlı,* 1954b: 40, 1966: 55; *Tolun,* 1962: 22; *Canuti* et al., 1970: 25).

The Devonian in East Anatolia (Fig. 7)

The Devonian sediments so abundant in Armenia seem to reach into Turkey at the foot of the Ağri (Mount Ararat) near Dogubeyazıt (*Blumenthal,* 1958: 253, 268).

Summary

The Devonian system is considerably more widespread than the lower Paleozoic, which is particularly true for Middle Anatolia (p. 20).

One of the symptoms of the Caledonian tectogenesis was a considerable regression at the Silurian-Devonian boundary. A continuity of marine deposits beyond that boundary in Turkey is known so far only on Kocaeli and in the Antitaurus.

Otherwise, the Emsian stage or Upper Devonian sediments are transgressive in many places. It is not clear whether the entire area had been flooded at the end of the Devonian period. If islands had remained in Middle Anatolia, their extent has not been so large as to impair the exchange of fauna between the northwestern and southern parts of Anatolia.

D e v o n i a n

	Bosporus and Western Kocaeli (Kaya 1973)	Eastern Kocaeli (Haas 1968 a)		Bartın (Charles 1933, Tokay 1954/55)	Sultan Dağ (Haude 1972)	Antitaurus (Özgül et al. 1973)	Amanos Dağ (El Ishmawi 1972, Lahner 1972)
Overlying	Baltalimanı Formation	Lower Carboniferous	Thracian series	Lower Carboniferous	Lower Carboniferous	Lower Carboniferous	Calcareous sandstone +
Upper Devonian: Famennian	110 m Büyükada Formation: Küçükyalı, Ayineburnu	100 m Denizli Formation	Tuzla series	Massive dolomitic limestone	70 m Engili quartzite, Conglomeratic quartzite	500 m Gümüşali Formation	10-400 m marl +
Upper Devonian: Frasnian	Yürükali; Nodular			1000 m	100 m Black shale + crinoid lms.	Sandstone + shale + reef limestone	shale + Hasanbeyli
Middle Devonian: Givetian	limestone + lydite; Bostancı	Gray and reddish nodular limestone		Bedded	± 1000 m	1700 m Şafak Tepesi Form. Massive limestone + dol.	limestone
Middle Devonian: Eifelian	60 m Içerenköy Form. Marl	200 m Gebze Form. Arenaceous marl		limestone	Greenish	400 m Ayı Tepesi	White and Formation
Lower Devonian: Emsian	50 m Kozyatağı Form. Marl + limestone	80 m Dede Formation. Nodular marl	Marmara series	400 m Glauconitic bedded lms.	gray	Formation	reddish quartzite
	400 m Kartal Formation	120 m Kurtdoğmuş F. Calcareous graywacke + sandstone		Conglomeratic quartzite	arenaceous	Sandstone + marl +	
Lower Devonian: Siegenian	Graywacke, shale + limestone	100 m Kartal Form. Arenaceous marl			shale	limestone	
Lower Devonian: Gedinnian	100 m Istinye Form. Bedded limestone + marl	50 m Soğanlı Form. Nodular limestone + marl				Yukarıyayla Formation	
Underlying	Dolayba limestone	200 m Kireçhane Formation. Black bedded limestone	Akviran series	? Ordovician	Ordovician	Yukarıyayla Formation ↑	Silurian

As early as the end of the Silurian period, the maximum depth of the sea had started to shift from South to Northwest Anatolia. In the Late Devonian, the result becomes quite evident from the facies difference exhibited by the nodular limestone of Kocaeli and the coral marl of the Antitaurus.

The shores of the Tethys were displaced in a correspondent manner. The northern coastline was transgressive and progressed towards the Black Sea. The southern coastline receded relative to the Silurian; no Devonian marine sediments are known in Syria and Jordania.

The thickness of the Devonian strata reaches approximately 1500 m. Diabase sheets and tuffs are more abundant than previously, particularly in the Antitaurus. Nevertheless, as a whole, the sequence has to be regarded as miogeosynclinal.

The marine connections stayed the same. The Turkish Devonian fauna is most closely related to that of Central and Southern Europe. But there is no full agreement; a number of local species give the region a special character. Some other species are linked with those present in the Ural Mountains and Central Asia (*Paeckelmann,* 1938: 185; *E.* and *H. Flügel,* 1961: 397; *Haas,* 1968b: 193).

Chapter 9 Carboniferous

In 1822, bituminous coal was discovered at Zonguldak-Amasra. *Schlehan* showed in 1852 that it belongs to the Carboniferous system. Not much later, in 1854, *De Tchihatcheff* collected the first Carboniferous fossils in the Antitaurus.

Descriptions of Carboniferous marine fauna have been written by *Frech* (1916), *Charles* (1933), *Delépine* (1933), *H. Flügel* and *Kıratlıoğlu* (1956), *Ünsalaner* (1958), *Wagner-Gentis* (1958), and *Kaya* (1973); of foraminifera and calcareous algae by *Güvenç* (1965, 1966); of macrofloras by *Zeiller* (1899), *Ralli* (1933), *Jongmans* (1939), and *Egemen* (1959). Literature on sporomorphs has been cited by *Artüz* (1963), *Agralı* and *Konyalı* (1969), and *Yahsıman* (1972). *Chaput* (1936: 239), *Heritsch* (1939a), *H. Flügel* (1964, 1971), and *Wolfart* (1967b) have published summaries.

The Carboniferous of Istanbul and Kocaeli (Fig. 8)

Based on fossil findings by *Yalcınlar* (1951), the Thracian series, which *Paeckelmann* (1938: 39) regarded as an Upper Devonian facies, was placed into the Carboniferous system. The sequence has been described by *Haas* (1968a: 230) and *Kaya* (1969, 1971, 1973).

The depth of the sea which had gradually increased since the Middle Devonian epoch, reached its maximum at the beginning of the Early Carboniferous. The Devonian nodular limestone has been replaced by chert in the Tournaisian stage. Then, a strong influx of clastic material created a thick graywacke flysch sequence. The area of denudation was still situated in the north and northeast of Kocaeli, as shown by current markings. Several conglomeratic sequences with pebbles of vein quartz, quartzite, and granite are indicators of its source. The proximity of a coastline is also demonstrated by numerous plant remains which are present in the sediments in association with foraminifera and goniatites. During a short period of shallowing, coral-brachiopod coquina was formed. The youngest graywackes reach as far as the basal part of the Upper Carboniferous. They are particularly rich in plant remains and may have been deposited in a delta. It thus becomes evident that the Carboniferous profile ended with a regression at about the end of the Namurian stage. A diabase sheet is the only volcanic intercalation.

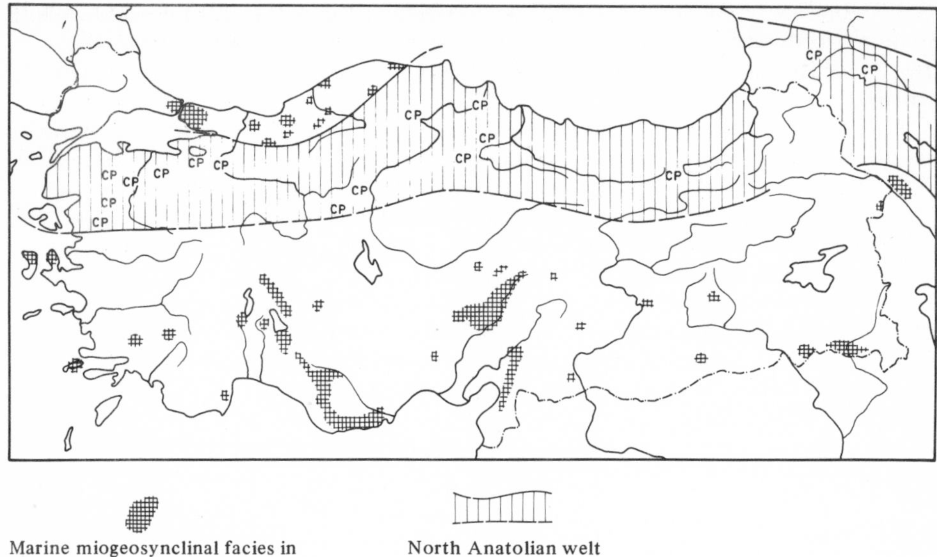

Fig. 8 Distribution, facies and paleogeography of the Carboniferous in Turkey

The Carboniferous of Zonguldak-Amasra-Azdavay (Fig. 8, 9)

The geological relations of the Turkish coal district have been clarified by *Charles* (1933, 1954), *Arni* (1939b), *Grancy* (1939), *Tokay* (1952, 1954/55, 1962), *Patijn* (1953/54), and *Jongmans* (1956), who relied on the investigations by *Schlehan* (1852) and *Ralli* (1895/96).

The Lower Carboniferous as well as the Upper Devonian sedimentation of this area has taken place in shallower water and with less detrital material than at Istanbul. Thick beds of coral-brachiopod limestone reach from the Middle Devonian to the Visean stage. They are overlain by dark, fossil-rich shale with ostracodes, mollusks, and goniatites, which probably comprise the upper Visean and the lowermost Namurian stage (*Wagner-Gentis,* 1958; *Petrascheck,* 1955).

Then, the sea started to recede; the productive coal measures in the Upper Carboniferous are mostly limnic (except a Lingula horizon in the Westphalian A; *Loboziak* and *Dil,* 1973). They are partly composed of arkose and graywacke, and partly of conglomerates in which sillimanite-bearing quartz grains and pebbles of metamorphic and magmatic rocks, as quartz porphyry, more rarely dacite and andesite can be found (*Bayramgıl,* 1951). Nothing is known about the location of the area of denudation, but it appears that at least part of the clastic material was derived from the Pontian land (*Brinkmann,* 1974: 64). *Zijlstra* (1952) and *Petrascheck* (1955) draw attention to layers of tuff, washout and coal rubble; *Artüz* (1974) investigated the petrography of the coal.

About fifty coal seams are known, thirty of which are mineable. The first seams appear in the late Namurian stage. The early Westphalian stage is richest in coal. With the Stephanian stage, the rock color changes from gray to red. and coal seams disappear. This red series, which extends as far as the Permian system, is widely developed in the eastern part of the coal area between Azdavay, Inebolu, and Cide (*Grancy,* 1939: 80; *Fratschner,* 1954: 212; *Altınlı,* 1956:16). *Stchépinsky* reports Callipteridium pteridium in these rocks (1946: Pl. 5, Fig. 1).

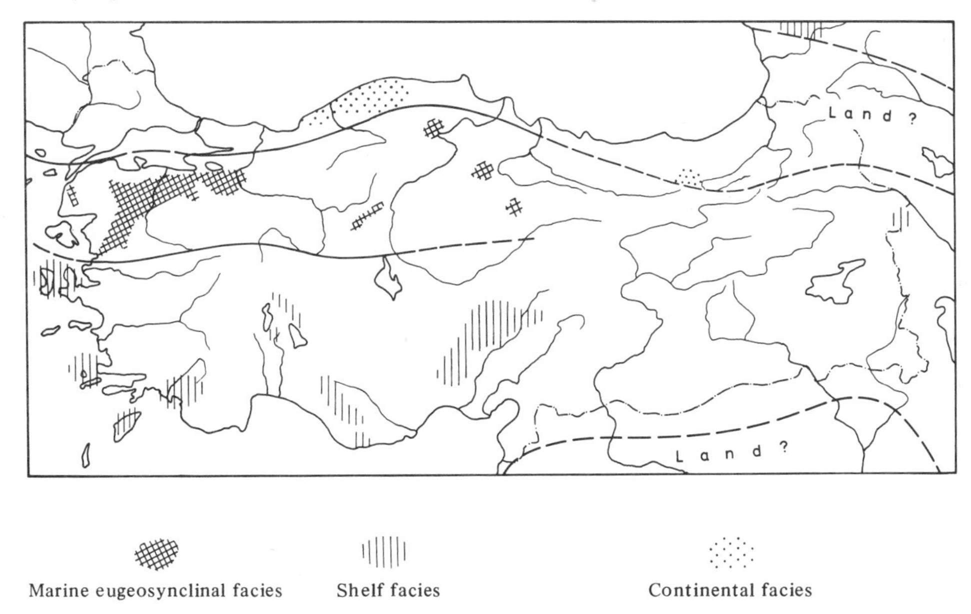

Fig. 9 Distribution, facies and paleogeography of the Upper Carboniferous in Turkey

The Carboniferous in West Anatolia (Fig. 8, 9)

At Chios, Carboniferous sequences of two different facies are in tectonic contact (*Besenecker* et al., 1968: 129, 134). It is likely that the same is true for Karaburun. Fossils of the Visean (*H. Gümüs,* 1971: 7) as well as of the Upper Carboniferous (*Garrasi* and *Weitschat,* pers. com.) have been found in limestones with graywacke.

The Carboniferous in Middle Anatolia (Fig. 8, 9)

In a zone which extends from the Biga Peninsula in the west beyond Ankara to the east, Lower Paleozoic sediments are almost absent; Upper Paleozoic strata are resting directly on metamorphic rocks. This zone was dry land at the beginning of the Carboniferous period. During the Visean stage it was partially flooded for a short time. However, only a few remains of the dark limestone deposited during that period, are left. Most of the limestone beds became eroded rapidly. They are found today as exotic blocks in the graywacke flysch of the Orhanlar formation, together with fragments of metamorphites, magmatites, Devonian, Upper Carboniferous, and Lower Permian limestones (*Chaput,* 1936: 261; *Erol,* 1958: 7, 1968: 17; *Brinkmann,* 1971c; *Alp,* 1972: 32). These strata – frequently diabase-bearing – are overlying the crystalline base from Kaz Dağ in the west to at least as far as Amasya in the east. Farther east, at Bayburt, there are outcrops of multicolored clastic rocks with comparable stratigraphical relations, which may be continental equivalents of the marine Orhanlar graywacke (*Ketin,* 1951: 116).

The Carboniferous in Southwest and South Anatolia (Fig. 8, 9)

Quotations of Lower Carboniferous corals at Muğla (*Önay,* 1949: 364) and Carboniferous calamites and fusulines at Denizli (*Yalcınlar,* 1963: 19) have not been substantiated.

Outcrops of rocks of definitely Carboniferous age are found south of Denizli (*Altinli,* 1955: 15), at the Eğridir Gölü (*Ünsalaner,* 1958) in the Sultan Dağ (*Haude,* 1972: 415), and at Fethiye (*De Graciansky,* 1972: 103). These exposures indicate that the Carboniferous in the western Taurus is composed of more than 500 m of graywacke, shale, limestone, and locally of tuffite. This sequence represents Tournaisian to Moscovian stages.

In contrast to the thick, lithologically varied development in the western Taurus, in the eastern Taurus and the Antitaurus, at Malatya (*Stchépinsky,* 1944: 96), and perhaps in the Bitlis Mountains, the system is represented by an only 100 to 200 m-thick sequence of dark limestone beds with some intercalated, fine-clastic material. The strata are rich in foraminifera, corals, brachiopods, gastropods, and calcareous algae. They comprise the entire Carboniferous system from the Tournaisian to the Gshelian. Lower Carboniferous sediments have been reported from several places (*Frech,* 1916: 209; *Blumenthal,* 1941: 59, 64, 1947b: 54, 1951: 27, 43; *H. Flügel* and *Kıratlıoğlu,* 1956; *Ünsalaner,* 1958; *Vaché,* 1964: 93; *Güvenç,* 1965: 33, 1966). Because of its small thickness, the Upper Carboniferous has long been overlooked (*Güvenç,* 1965: 38).

The Carboniferous in Southeast Anatolia (Fig. 8, 9)

In the Amanos Dağ region, the Upper Devonian continues conformably into the Lower Carboniferous (*El Ishmawi,* 1972: 45; *Lahner,* 1972: 74). It consists of coral limestone alternating with sandstone and shale. Upper Carboniferous sediments are not known with certainty. In Southeast Anatolia, thin sequences of Lower Carboniferous deposits are reported from several places, such as Harbol and the Ricgar Dağ (*S. Türkünal,* 1953: 9; *Altınlı,* 1954a: 10). In other areas, such as Hazro (p. 34) and south of Hakkari (*Altınlı,* 1954b: 41, 1966: 56), the Carboniferous system seems to be missing altogether. It is possible that during the late Carboniferous period, southeast Anatolia was dry land.

The Carboniferous in East Anatolia (Fig. 8, 9)

A massive dark limestone south of Sivas contains upper Visean algae (unpub. data).

The calcareous Carboniferous of Armenia is probably also represented in outcrops in Turkey around the Ağrı. *Blumenthal* (1958: 268), *Altınlı* (1966: 42), and *Batum* (1969: 14) mention several occurrences.

Summary

The Carboniferous system in Turkey is dominated by strong differences in facies. As in Central Europe, the profile of Northwest Anatolia can be subdivided into two complexes. The Lower Carboniferous is of marine origin and developed partly as Culm, and partly as Carboniferous limestone. The Upper Carboniferous is continental and contains coal beds. In the southern part of Turkey and in the remainder of the Near East, the entire Carboniferous system is composed of calcareous sediments deposited in a shallow sea. Isolated basic or acid volcanite and tuff are intercalated here and there.

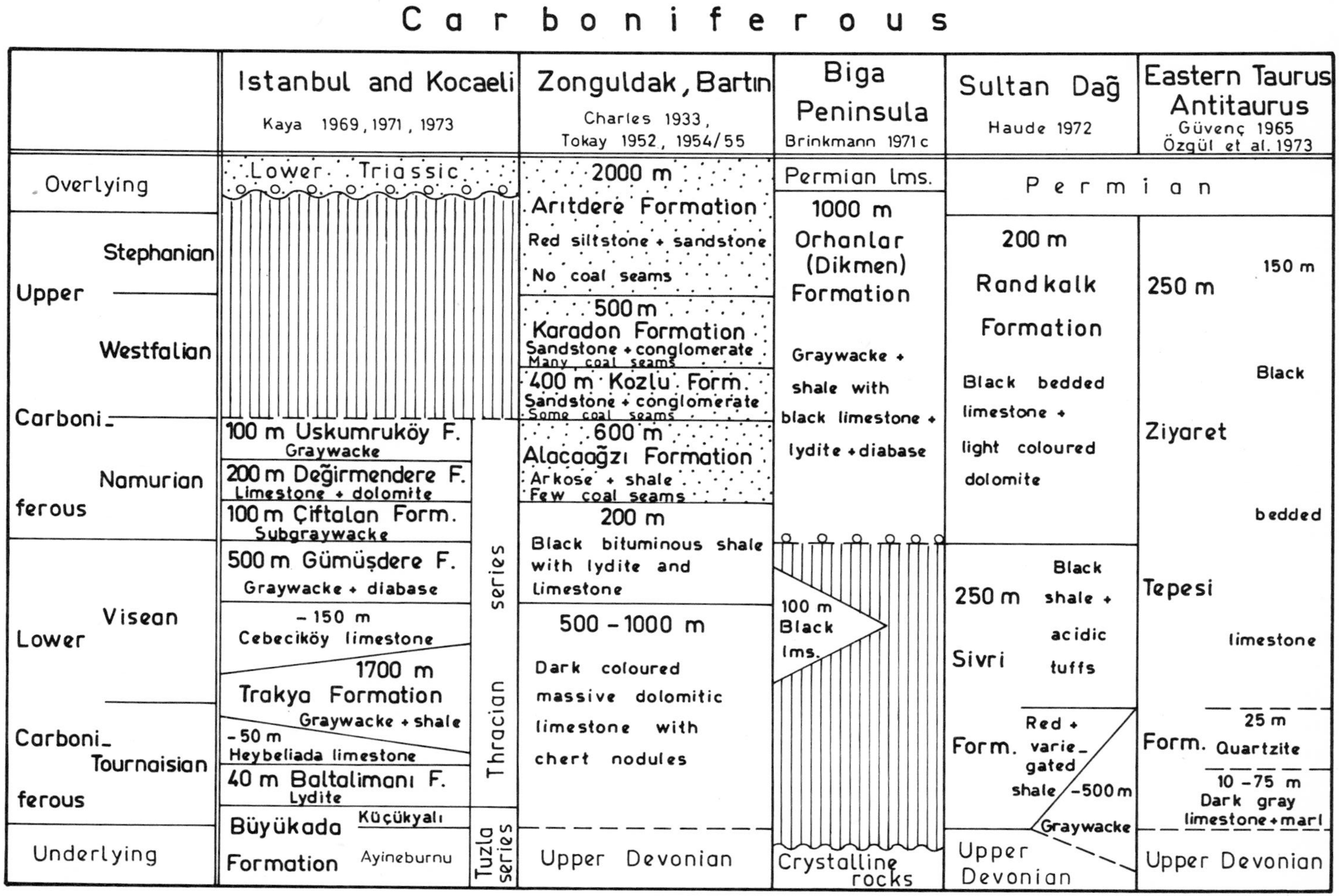

Carboniferous

	Istanbul and Kocaeli Kaya 1969, 1971, 1973	Zonguldak, Bartın Charles 1933, Tokay 1952, 1954/55	Biga Peninsula Brinkmann 1971c	Sultan Dağ Haude 1972	Eastern Taurus Antitaurus Güvenç 1965 Özgül et al. 1973
Overlying	Lower Triassic	2000 m Arıtdere Formation Red siltstone + sandstone No coal seams	Permian lms.	Permian	Permian
Upper Carboniferous: Stephanian			1000 m Orhanlar (Dikmen) Formation Graywacke + shale with black limestone + lydite + diabase	200 m Randkalk Formation Black bedded limestone + light coloured dolomite	250 m, 150 m Ziyaret Tepesi Form. Black bedded limestone
Upper Carboniferous: Westfalian		500 m Karadon Formation Sandstone + conglomerate Many coal seams 400 m Kozlu Form. Sandstone + conglomerate Some coal seams			
Upper Carboniferous: Namurian	100 m Uskumruköy F. Graywacke 200 m Değirmendere F. Limestone + dolomite 100 m Çiftalan Form. Subgraywacke	600 m Alacaağzı Formation Arkose + shale Few coal seams 200 m Black bituminous shale with lydite and limestone			
Lower Carboniferous: Visean	Thracian series: 500 m Gümüşdere F. Graywacke + diabase -150 m Cebeciköy limestone 1700 m Trakya Formation Graywacke + shale	500 – 1000 m Dark coloured massive dolomitic limestone with chert nodules	100 m Black lms.	250 m Sivri Form. Black shale + acidic tuffs	
Lower Carboniferous: Tournaisian	-50 m Heybeliada limestone 40 m Baltalimanı F. Lydite Tuzla series: Büyükada Formation Küçükyalı Ayineburnu			Red + variegated shale -500 m Graywacke	25 m Quartzite 10 – 75 m Dark gray limestone + marl
Underlying		Upper Devonian	Crystalline rocks	Upper Devonian	Upper Devonian

In the Lower Carboniferous, going from north to south, the following paleogeographic and tectonic zones can be distinguished:

1. The Pontian land was situated in the Black Sea.
2. South of it extended the Tethyan geosynclinal sea. Epirogenic movements at the beginning of the Carboniferous subdivided the sea into two straits, a northern narrower strait, and a southern wider one. In both straits the subsidence and sedimentation continued from the Devonian period to the Carboniferous without interruption. The Carboniferous system in the northern strait, however, reached thicknesses that were 5 to 10 times greater than those in the southern strait. The geanticline in between – the *North Anatolian welt* (*Brinkmann,* 1968: 112) – may have been wholly or partly dry land.
3. In the south, at the shores of the Afro-Arabian craton, the Lower Carboniferous developed transgressively. Marine sediments can be found as far south as central Syria; they indicate the proximity of the mainland, however, because ot their sandy character and their content of plant debris.

At the beginning of the Upper Carboniferous, an epirogenic change occurred, which manifested itself in the following manner:

1. The Pontian land remained unchanged.
2. A thick series of clastic material, which originated partly from the Pontian land, and probably partly from the south, filled up the northern strait. Continued subsidence caused the accumulation of a bituminous coal formation, only a small part of which is situated in present-day Anatolia. The Upper Carboniferous of Northwest Anatolia extends into the Black Sea at Ereğli in the west and at Inebolu in the east. The North Anatolian welt subsided, and a eugeosynclinal flysch trough took its place until the Permian period. As in the Early Carboniferous, the southern strait remained a shallow sea with slow sedimentation rate.
3. The southern coast of the Tethys was regressive during the Upper Carboniferous period. Southeast Anatolia appears to have temporarily emerged from the sea.

In spite of their narrowing, however, the straits remained open to the east and west. As in the Devonian, the marine fauna is related with that of Europe, but ties with East Asia have also been established (*H. Flügel* and *Kıratlıoğlu,* 1956). The flora of the bituminous coal beds of Zonguldak-Amasra shows a purely Central European character.

Chapter 10 Permian

In Turkey, the first localities of Permian fusulinid limestone were reported by *Texier* (1835) at Tarsus and by *Neumayr* (1887) at Balya. But the vast extent of the system was not known until studies by *Philippson* (1910–15) and *Blumenthal* (1941–63). *Erk* (1942), *Ciry* (1941–43), *Dessauvagie* and *Dağer* (1963), *Sellier* and *Dessauvagie* (1965), *Güvenç* (1965, 1967), and *Skinner* (1969) have described certain foraminifera groups, *Enderle* (1901), *Heritsch* (1939b), *Metz* (1939a), *H. Flügel* (1955c, d), *Aygen* (1956) studied parts of the remaining fauna. *Bilgütay* (1960b) and *Güvenç* (1972) have written on the calcareous algae. Summaries on the Permian in Turkey have been prepared by *Chaput* (1936: 240), *Heritsch* (1939a), *Sellier* and *Dessauvagie* (1965: 28), *H. Flügel* (1964, 1971), *Wolfart* (1967b), and *Kahler* (1974: 81).

The Permian in North Anatolia (Fig. 10)

In the coastal area of the Black Sea, the Permian and the Upper Carboniferous systems show continental facies. The gray, coal-bearing sediments of the Westphalian stage pass into a barren red series. Their lower part belongs to the Stephanian stage (p. 29). Except for one Walchia locality (*Wedding,* 1970: 43), the upper part is non-fossiliferous. It is therefore uncertain how far it reaches into the Permian. Fine-bedded siltstone and sandstone are predominant; conglomerates of quartz and chert pebbles are limited to the lower part (*Grancy,* 1939: 80; *Fratschner,* 1954: 212; *Altınlı,* 1956: 16). The red series is missing at Zonguldak, which is probably due to later erosion. It appears near Bartin and increases in thickness towards the east, in the area between Cide and Azdavay southwest of Inebolu.

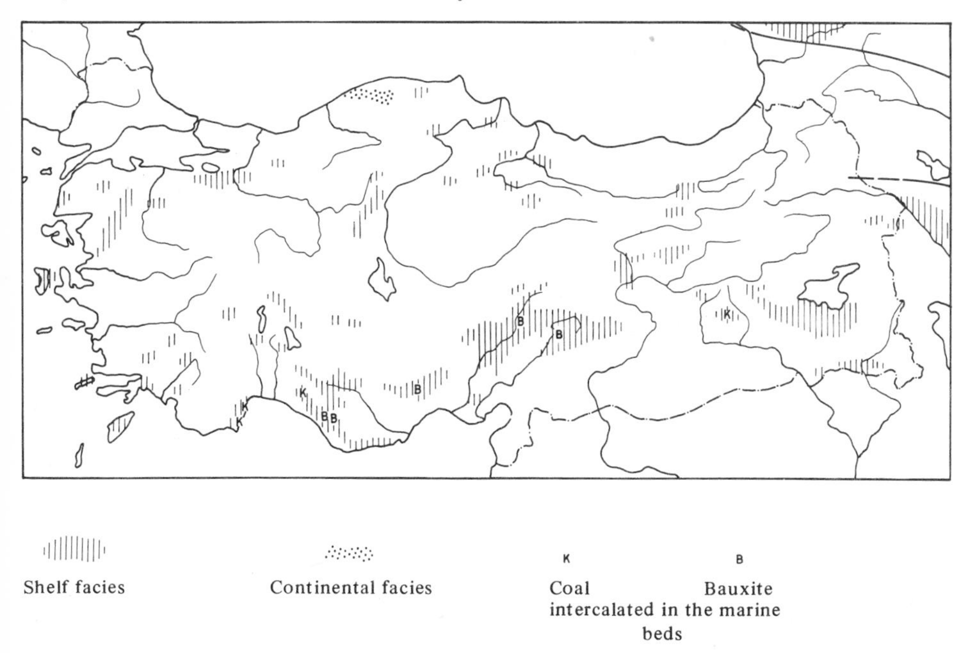

Fig. 10 Distribution, facies and paleogeography of the Middle/Upper Permian in Turkey

It is probable that the 700 m-thick layers of multicolored arkose, sandstone, and conglomerates, which *Ketin* (1951: 116) is describing from Bayburt in Northeast Anatolia, belong to this Upper Carboniferous/Permian series. They are overlain by Permian Fusulina limestone.

The Permian in Middle Anatolia (Fig. 10)

As in the Upper Carboniferous, marine sediments are predominant in this area. The flysch sequence of the Orhanlar formation continues into the Lower Permian. Its youngest parts contain autochthonous Permian fossils and blocks of Lower Permian limestone (*Boccaletti* et al., 1966a: 487; *Erol,* 1968: 17; *Akartuna,* 1968: 36; *Brinkmann,* 1971c: 58). The gray-wackes are distinctly overlain by bituminous, gray to black limestone beds. They are rich in microforaminifera, fusulinids, bryozoans, brachiopods, crinoids, and calcareous algae; some beds are entirely composed of the skeletal remains of these groups. Oolitic limestone, marl, light-colored quartz sandstone, quartzite, and occasionally red arkose are intercalated (*Erk,* 1942: 34; *Blumenthal,* 1950: 28, 1958: 267; *Aygen,* 1956: 71; *Afshar,* 1965: 33; *Altınlı,* 1966: 42; *Lys,* 1971/72).

The Permian in South and Southeast Anatolia (Fig. 10)

The deposition of beds of dark-colored limestone with fusulinids and calcareous algae, which in Middle Anatolia was limited to the later Permian period, continued in the Sultan Dağ, in the eastern Taurus and the Antitaurus, and possibly also near Malatya, and in the Bitlis Mountains from the Carboniferous through almost the entire Permian (*Metz,* 1955; *Tolun,* 1962: 213; *Güvenç,* 1965: 57; *Haude,* 1972: 416). Intercalations of various nature interrupt the rather uniform sequence. Reddish, pisolithic algal limestone must have formed in very shallow water (*Blumenthal,* 1960/63: 617). In several places, bauxite lenses are found (*Wippern,* 1965), such as at Alanya (*De Peyronnet,* 1971: 96; *Argyriadis,* 1974), in the Bolkar Dağ (*Blumenthal,* 1955: 53, 160), and in the Antitaurus (*Blumenthal,* 1944c). The diaspore-corundum rocks of the southern Menderes Massif (p. 7; *Önay,* 1949) may in part belong to the Permian system. Some profiles contain multicolored arkose, such as at Fethiye (*De Graciansky,* 1972: 123) and south of the Beyşehir Gölü (*Blumenthal,* 1947a: 181; *Brunn* et al., 1972a: 539). In Southwest Anatolia at Milas, the Permian is entirely composed of clastic red beds (*Brinkmann,* 1967: 5) which pass into marine sediments towards the east.

Along the southern coast of Anatolia and in Southeast Anatolia the Permian appears to have been deposited near shore. At Alanya (*Blümel,* 1969: 429) and in Southeast Anatolia (*G. C. Schmidt,* 1964: 106; *Ağralı* and *Akyol,* 1967; *Canuti* et al., 1970: 27) Upper Permian sediments are overlapping pre-Carboniferous rocks. At Fethiye, southwest of Antalya, at Alanya, Akseki (*Nebert,* 1964a) and at Hazro northeast of Diyarbakır, thin coal seams are intercalated with Permian marine sediments. Near Hazro a remarkable late Permian flora has been found (*Wagner,* 1962; *Ağralı* and *Akyol,* 1967). It consists mostly of species of the Cathaysian realm, some forms of the Angara and Gondwana realms, but only one single European form. According to *G. C. Schmidt* (1964: 108), the same flora is present at Harbol near the Iraqi border.

As the facies relations show, South and Southeast Anatolia belonged at that time to a shallow sea, which was bordered by land in the south. The shore-line was shifting considerably; at times, parts of southern Turkey were dry land. The volcanic activity was negligible. The Permian of Nif near Fethiye contains diabase sheets (*De Graciansky,* 1972: 130). At Sandıklı south of Afyon, layers of porphyroids in association with Verrucano-like conglomerates have been reported (*Brunn* et al., 1972a: 520; *Öngür,* 1973). *Paréjas* (1943) placed them into the Permian, maybe they are indeed geotectonically analogous to the Rotliegende in Central Europe.

Summary

During the Permian, the distribution of land and sea in the Near East was almost completely reversed to that of today. Anatolia was nearly entirely covered by water. The northern and southern shorelines of that sea coincided approximately with the present coastlines, only in a reversed sense. The Pontian land to which also belonged a narrow strip of North Anatolia, was situated in the Black Sea. In the south, Israel and Jordania remained above sea level for almost the entire Permian period. Northern Syria, and South and Southeast Anatolia were flooded periodically; marine limestones interfingered with continental sediments.

The Permian Tethyan geosyncline was of simple shape. Until the end of the Early Permian epoch, the Orhanlar flysch trough was a special unit. During the Late Permian, the facies and thickness of the sediments throughout Anatolia were rather uniform. Basic volcanites are present in the flysch trough, whereas they are quite rare in the calcareous sequences.

At the beginning of the Permian, the late Paleozoic regression seems to have been at its peak. From then on, a reversal took place. Upper Permian marine sediments are overlapping in the north and the south of Turkey (p. 33, 34). Shortly before the close of the Permian, a general shrinking process set in, inasmuch as equivalents of the late Djulfian stage so far have not been found in Turkey (*Güvenç,* 1965: 88).

The Anatolian Tethys was inhabited by organisms of glôbal relationship. The foraminifera, corals, brachiopods, and calcareous algae are close to those in southern Europe and southern Asia, but also to those in North America, Indonesia, and East Asia (*Heritsch,* 1939b: 185). Such an extensive exchange requires open waterways. Indeed, the Permian Tethys was transgressive not only in Anatolia, but in the remaining Near East (*Wolfart,* 1967b: 177). The continental flora is only preserved at Hazro. It is most closely related to that of western Iraq (*Čtyroký,* 1973) and is in sharp contrast to the purely Central European Upper Carboniferous flora of Zonguldak. The boundaries of the floral realms in Asia must have shifted considerably from the Carboniferous to the Permian period.

Permian

	Bartın, Inebolu (Grancy 1939, Altınlı 1956)	Biga Peninsula (Brinkmann 1971 c.)	E. Taurus, Antitaurus (Güvenç 1965)	Hazro (Tolun 1962, Ağralı & Akyol 1967)	Cizre (G.C. Schmidt 1964)
Overlying	Liassic	U. Triassic	L./M. Triassic	L. Triassic	L. Triassic
Djulfian (= Pamir, Tatar)			200-500 m	50 m Marly limestone	300 m
Murgabian		400 m Permian	Permian limestone	90 m Limestone	Harbol Formation
Kasanian / Kubergandian		limestone Dark gray	Dark gray bedded	100 m Sandstone + shale; 25 m Limestone	Dark gray bedded limestone
Artinskian	2000 m Aritdere Formation	bedded limestone with marl + sandstone	limestone with sandstone, arkose,	100 m Sandstone + shale Few coal seams	with sandstone + shale
Sakmarian	Red siltstone + sandstone	1000 m	bauxite and coal beds intercalated		
Asselian		Orhanlar Graywacke			
Underlying			Upper Carbon.	U. Devonian	L.Carbon.

Chapter 11 Variscan Tectogenesis and Metamorphism

The importance of Variscan crustal movements in Turkey has first been recognized by *W. Penck* (1919) and later emphasized by *Paeckelmann* (1938). *Ketin* (1959b) and *H. Flügel* (1964) have written on the extent of the movements in Turkey and the Near East.

The Variscan Tectogenesis in North Anatolia (Fig. 10–13)

The Paleozoic rocks in the Northwest Anatolian coastal area show generally Variscan folding. Near Istanbul the fold axes on either side of the Bosporus are mostly north-trending (*Kaya*, 1969:170, 1971: 185). It is possible that their direction was influenced by older structures which originated in the Caledonian era. On Kocaeli, there are either 120–140° or 30–50° trending anticlines and synclines, which are dissected by longitudinal and diagonal faults. Medium dip angles and a moderate north vergence are the rule. The cleavage is weakly developed and is steeply dipping toward the south (*Paeckelmann*, 1938: 160; *Haas*, 1968a: 183).

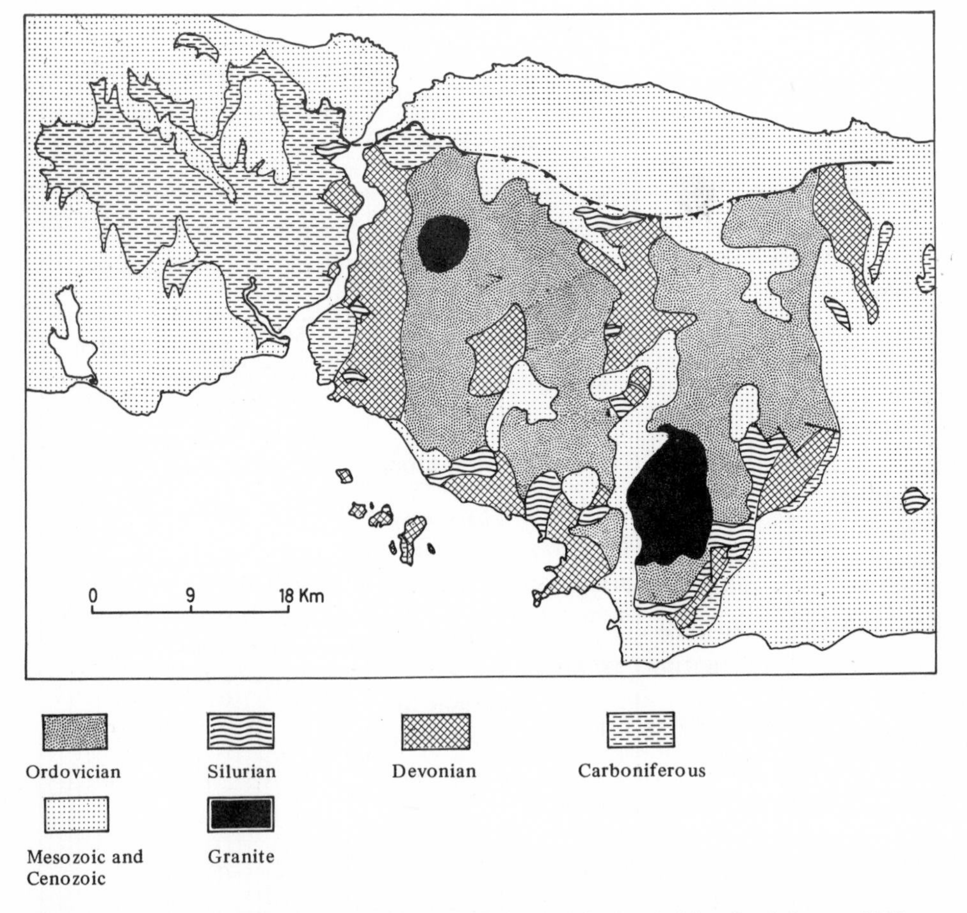

Fig. 11 Paleozoic area of Istanbul and of Kocaeli (Bithynian) Peninsula (after *Paeckelmann*, 1938, *Haas*, 1968, *Kaya*, 1971, *Ketin*, 1971, and others)

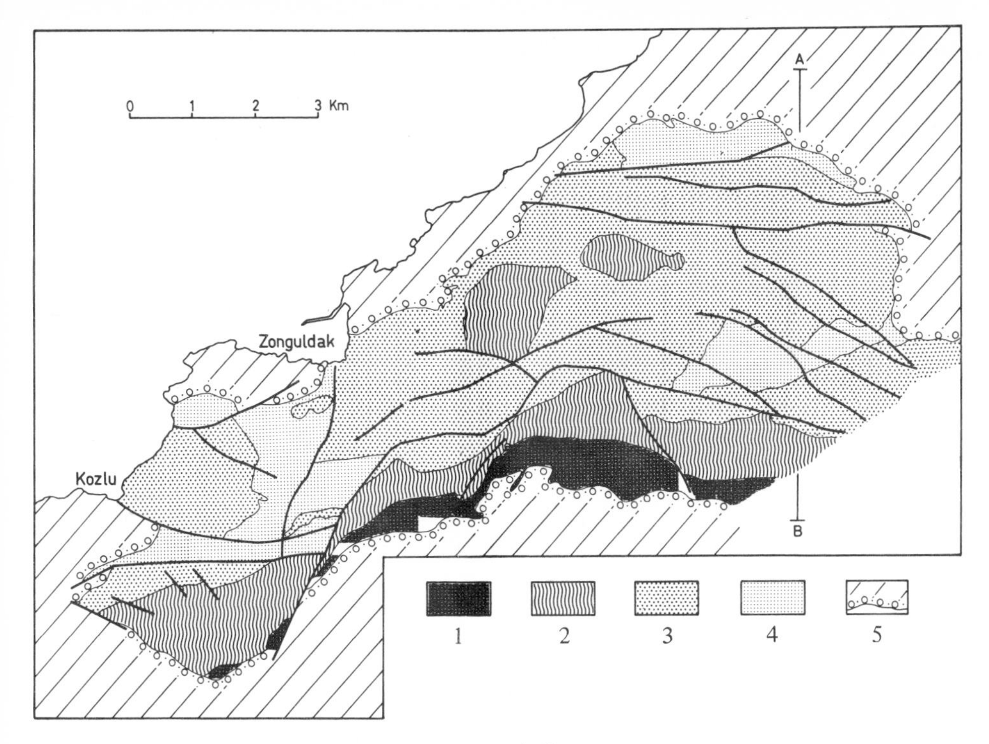

Fig. 12 Coal district of Zonguldak (after *Patijn,* 1953/54)

1 = Lower Carboniferous (Vise beds), 2–4 = Upper Carboniferous (Alacaagzi, Kozlu and Karadon beds),

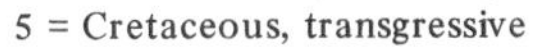
5 = Cretaceous, transgressive

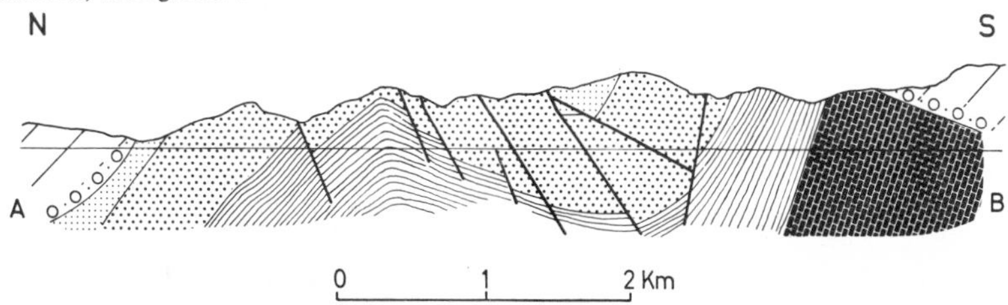

Fig. 13 Section across the coal district of Zonguldak (after *Patijn,* 1953/54). For explanation see fig. 12

In the Çam Dağ at Adapazarı the structure is similar. The major folding is east-trending, and the cross folding is north-trending. The angle of unconformity between the Paleozoic and the overlying rock formations reaches 90° (unpub. data).

In the bituminous coal district of Zonguldak-Amasra, the Variscan structures are difficult to ascertain, because they exhibit the same east-west trend as the Alpidic ones. The Paleozoic rocks have open symmetrical or slightly north-verging folds. Along the coast of the Black Sea, anticlines (at Ereğli and Bartin) and synclines (at Zonguldak and Amasra) are exposed (*Charles,* 1930: 176, 1954; *Tokay,* 1952: 62, 1962: 12; *Patijn,* 1953/54). The synclinal axis of Zonguldak dips west, because pre-Carboniferous rocks cut through the Cretaceous cover east of the town (*Altınlı,* 1951a: 159; *Ketin,* 1955a: 148).

Based on the following profiles, the age of mountain-building is late Variscan:

Istanbul-Kocaeli
(*Paeckelmann,* 1938,
Kaya, 1971; *Assereto,* 1972)

Triassic (Scythian)
? Upper Permian
Namurian
Lower Carboniferous
Devonian

Zonguldak-Amasra
(*Charles,* 1930; *Arni,* 1931;
Tokay, 1952, 1954/55)

Lower Cretaceous (Hauterivian-Aptian)
Permian (? Lower)
Upper Carboniferous
Lower Carboniferous
Devonian

Cide, west of Inebolu
(*Charles,* 1930; *Fratschner,* 1954;
Altınlı, 1956)

Jurassic (Liassic)
Permian (? Lower)
Upper Carboniferous

The Saalian and Palatian phases are most likely. Unfortunately, however, no precise information is available, due to the gaps.

According to *W. Penck* (1919: 30), the granites of Kocaeli in contact with lower Paleozoic rocks are considered to be of Variscan age, this the more so, because granite debris is present in the Triassic basal conglomerates (*Haas,* 1968a: 233). The biotite in the granite, however, shows a K-Ar age of 87.3 ∓ 3.0 m. y., according to *Bürküt* (1966: 208).

The Variscan Tectogenesis in Middle Anatolia (Fig. 14)

On the North Anatolian welt which traverses Middle and Northeast Anatolia in a westerly direction (p. 32), older Paleozoic sediments are almost missing. Generally Upper Carboniferous sediments are directly overlying the crystalline basement. Three explanations can be advanced to account for this phenomenon. Firstly, the pre-Carboniferous gap in the geologic record is original; which would imply that during earlier Paleozoic time, the welt separated the sea of Northwest Anatolia from the sea of South Anatolia as a kind of peninsula or island chain. The uniformity of the marine fauna and the lack of a marginal facies, however, argue against such an idea. Secondly, the lower Paleozoic strata covered the welt and became eroded prior to the Upper Carboniferous. Indeed, blocks of Devonian and Lower Carboniferous rocks are found in the Orhanlar graywacke (p. 29). Finally, it is possible that lower Paleozoic strata took part in the formation of the welt but have become unrecognizable by Variscan metamorphism.

Contradictions exist concerning the age of the crystalline rocks of the welt. On the one hand, unmetamorphosed lower Paleozoic rocks are overlying the phyllite and gneiss on (1) the the northern slope, such as in the Çam Dağ at Adazapari (unpub. data), at Mudurnu southwest of Bolu (*Abdüsselâmoğlu,* 1959: 81), in the Bolu Dağ (*Niehoff,* 1960), and at Daday south of Inebolu (*Grancy,* 1939: 86), as well as on (2) the axial part, such as at Amasya (*Alp,* 1972: 25). On the other hand, the radiometric age for the metamorphism of the Kaz Dag gneiss is late Variscan (p. 11). Also, some granite masses in the welt intruded during the Variscan orogeny: near Sögüt at the middle Sakarya (272 m. y.; *Çoğulu* and *Krummenacher,* 1967), in the Ulu Dağ (p. 11), in the Alacam Dağ southeast of Balikesir (intrusion 360 m. y., anatexis 330 m. y., according to *I. Yilmaz,* 1973), and at Gümüşhane (338–298 m. y.; *Delaloye* et al., 1972).

The North Anatolian welt, thus, is partly composed of pre-Variscan rocks which were metamorphosed in early Paleozoic or Cryptozoic time, and partly of metamorphites and magmatites which were formed during the Variscan tectogenesis. Only a more specific distinction between the two rock groups can decide as to the geotectonic role of the

North Anatolian welt. It either represents the central zone of the Variscan mountains in North Anatolia, folded and metamorphosed during the early Variscan phases, or it has to be considered as a purely epirogenic geanticline which rose at about the time of the Bretonian and once again at about the Sudetic phase. Later it subsided during the Upper Carboniferous epoch und turned into a flysch basin. At the end of the Permian period, it suffered a slight late Variscan warping, as evidenced by the overlap of Triassic strata in Northwest Anatolia (*Brinkmann,* 1971c: 62) and at Ankara (*Chaput,* 1936: 241 f.).

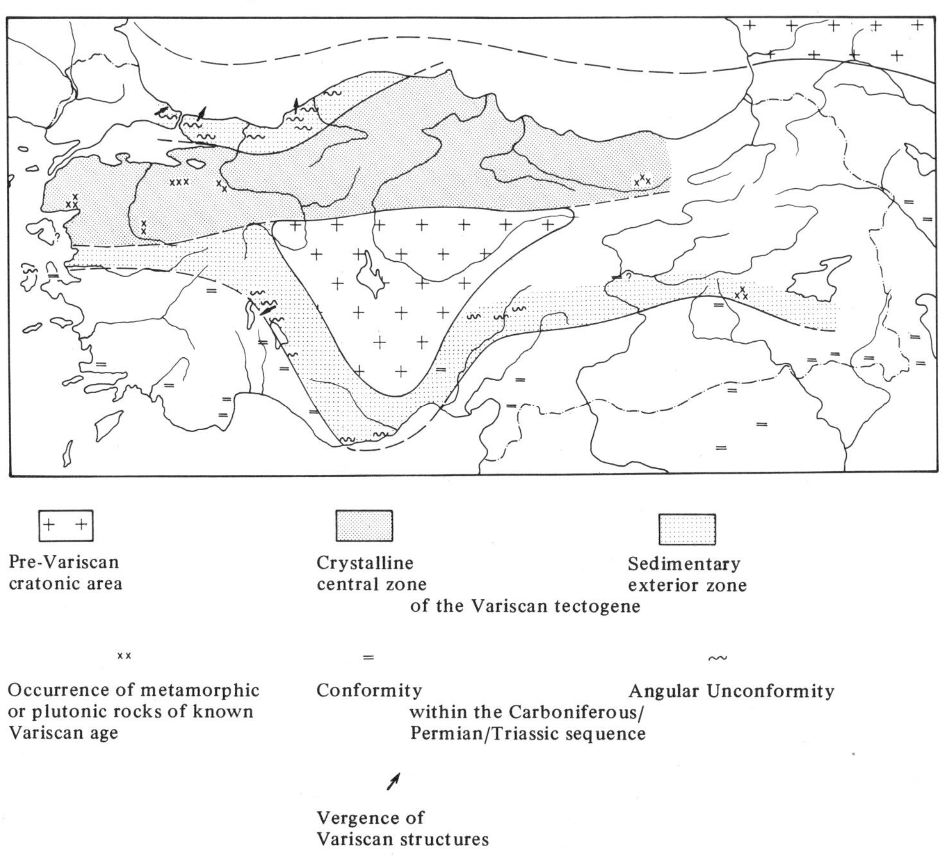

Fig. 14 Variscan mountain building in Turkey

Also the Sultan Dağ and its southwestern foreland as far as to the Beyşehir Gölü have been subjected to the Variscan tectogenesis. The sequence, which reaches conformably from the Cambrian to the Permian, has been thrusted into northwest-trending, southwest-verging folds. At the same time, parts of the area have been metamorphosed epizonally (*Haude,* 1972: 419). The cover beds start with the Anisian stage (*Monod,* 1967: 85). Here, too, therefore, the mountain-building was late Variscan.

Sultan Dag (*Haude,* 1972)	**Antitaurus** (*Özgül* et al., 1973)
Lower Triassic	Lower Triassic
Permian	Permian
Carboniferous	Lower Carboniferous
Devonian	Devonian

The Variscan Tectogenesis in South and Southeast Anatolia (Fig. 14)

Slight, late Variscan angular unconformities (i. e., within the late Paleozoic or between the Paleozoic and Mesozoic) have been frequently observed in the eastern Taurus at Anamur (*Işgüden,* 1971: 31; contrary to *Blümel,* 1969: 434) and Silifke (*Yalcınlar,* 1973: 109), in the Antitaurus (*Vohryska,* 1966: 101; *Özgül* et al., 1973: 99; unpub. data) and near Malatya (*Stchépinsky,* 1943: 117). Questionable intra-Carboniferous unconformities have been reported in some places too (*Wiesner,* 1968: 180; *Vaché,* 1964: 94). In the Bitlis Massif, the Permian system rests unconformably on the crystalline base and cuts off granite veins of probably Lower Carboniferous age (*Tolun,* 1953: 81; *O. Yilmaz,* 1971: 88).

The remainder of Anatolia, in particular the west coast area, the western Taurus, and Southeast Anatolia, stayed nearly or entirely untouched by the Variscan tectogenesis. On Karaburun, Triassic sediments are resting in slight unconformity on the Carboniferous (*Brinkmann* et al., 1972: 148). West of Beyşehir Gölü, the unconformity between the lower Paleozoic and Middle Triassic rocks is barely noticeable (*Dumont,* 1972). In Southeast Anatolia, the Permian is overlying Devonian or Ordovician strata (*S. Türkünal,* 1955: 53; *G. C. Schmidt,* 1964: 106, 109; *Canuti* et al., 1970: 27), however, without an obvious angular unconformity. The relationship can be explained by epirogenic movements which proceeded during the Variscan tectogenesis.

Summary (Fig. 14)

The map of the Variscan mountains in Turkey has been made under the assumption that the North Anatolian welt outlines the crystalline central zone of the Variscan tectogene. The folded Paleozoic strata between Istanbul and Inebolu, which probably continue on the bottom of the adjoining Black Sea, thus represent the northern sedimentary zone. It is characterized by a slight vergence to the north. The Variscan unconformities in the Sultan Dağ, eastern Taurus, Antitaurus, and the Bitlis Massif can be connected to a southern sedimentary zone. In this case, an arc results which surrounds the southern edge of the Inner Anatolian Massif (p. 10), and shows a vergence towards the south.

In this model, regional metamorphism and plutonism are diminishing toward the exterior, and the age of the main folding is decreasing in the same direction. The central zone received its tectonic structure during Carboniferous time. It had already subsided below the level of the sea before the external zones were folded during the Permian period.

Chapter 12 Triassic

The first Triassic fossil localities have been Balya in Northwest Anatolia (*Neumayr,* 1887) and soon thereafter the Kocaeli Peninsula (*Toula,* 1895). Major progress has been made by *Von Bukowski* (1892) and *Chaput* (1932). Parts of the marine invertebrate fauna have been investigated by *Bittner* (1892–96), *Toula* (1896), *Von Arthaber* (1915), *Brönnimann* et al. (1970), *Freneix* (1971/72), *Cuif* and *Fischer* (1974), and *Yurttaş-Özdemir* (1971, 1973). Summaries have been compiled by *Chaput* (1936: 241) and *Leuchs* (1939).

The Triassic in West Anatolia (Fig. 15)

So far, the most complete marine Triassic profile has been discovered on the Karaburun Peninsula (*Brinkmann* et al., 1972). Following a stratigraphic gap, which comprised at least the Permian, the sea advanced during the middle Scythian stage. At the beginning, it was shallow; during the Anisian stage it reached its greatest depth with the deposition of reddish ammonite limestone of the Hallstatt type and multicolored radiolarites. The Middle and Upper Triassic strata consist of the lithological equivalents of the Wetterstein limestone, the Raibl beds, and the Hauptdolomit (*Ott,* 1972). The Rhaetian stage is uncertain, but obviously, a gradual transition from the Triassic to the Jurassic system takes place. The Triassic rocks of Chios Island are closely related to those of Karaburun (*Besenecker* et al., 1968). In both places, keratophyre eruptions occur at the Scythian-Anisian boundary.

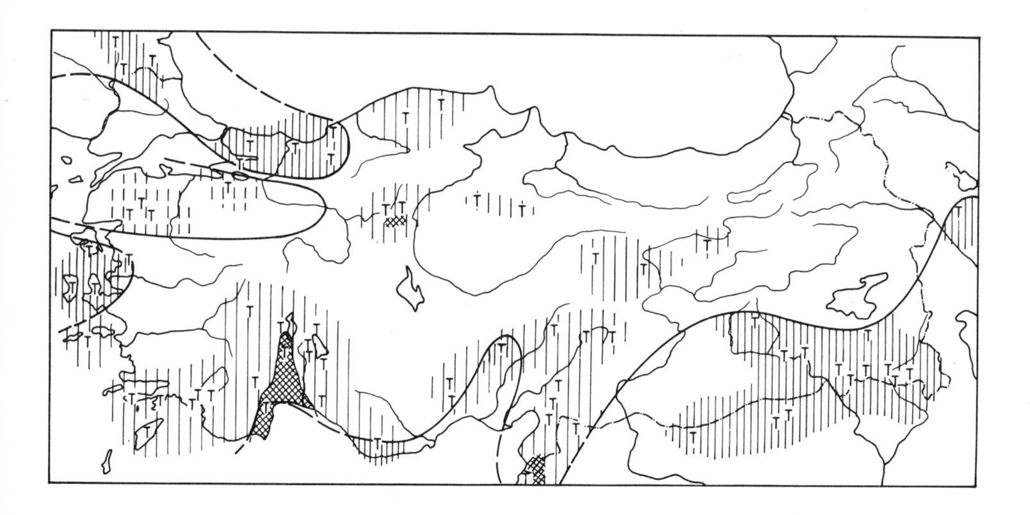

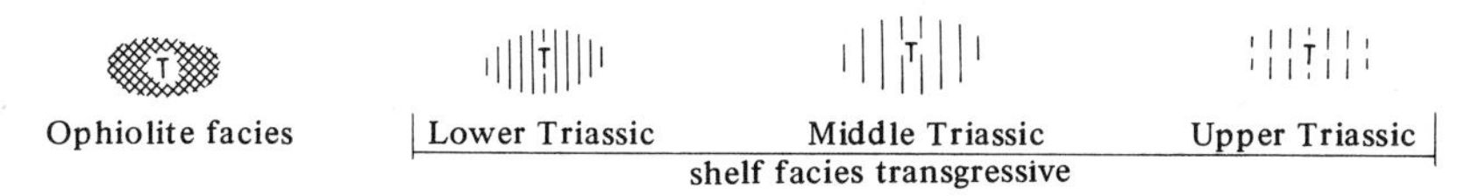

Fig. 15 Distribution, facies and paleogeography of the Triassic in Turkey

Farther inland from Izmir, this lithologically detailed sequence has not yet been found. However, dolomite marbles in the periphery of the Menderes Massif may represent metamorphosed Triassic sediments. *Dürr* (pers. com.), and *Wippern* (1965: 41), particularly regard the marble beds with diaspore-corundum lenses (p. 7) to be of Triassic age.

The Triassic in Northwest Anatolia (Fig. 15)

On Kocaeli Peninsula, the Paleozoic strata that were folded by the Variscan tectogenesis, are unconformably overlain by red sandstone with basal conglomerates. They represent a continental facies of the Scythian, and may also be of the uppermost Permian (*Assereto,* 1972: 437). Towards the top, they change into upper Scythian dolomitic limestones. At Sile these are succeeded – slightly unconformably (*Baykal,* 1942: 182) – by the Anisian to Carnian stages. The rocks consist of pink to gray, ammonite-bearing nodular limestone (*Von Arthaber,* 1915: 86; *Altınlı* et al., 1970a: 71; *Assereto,* 1972, 1974). The uppermost unit consists of shale and sandstone, which may reach into the Norian stage (*Yurttaş-Özdemir,* 1971, 1973).

The early termination of the profile is probably due to subsequent erosion, because farther east, the Triassic system, except for a few remains, has been completely removed. Red, conglomeratic sandstones, which are overlying the Paleozoic strata unconformably, are known in the Çam Dağ (unpub. data), the Bolu Dağ (*Niehoff,* 1960), and maybe at Zonguldak (*Arni,* 1931: 308). Only at Devrekâni south of Inebolu, Anisian limestone beds are preserved (*Blumenthal,* 1948: 109, 245).

In contrast to the nearly complete, mostly calcareous sequence on Karaburun, Chios, and Kocaeli, on the Biga Peninsula at Edremit and Balya, and near Bursa only late Triassic sediments in clastic facies are found (*Brinkmann,* 1971c: 59). According to *Zapfe* (1967: 15), the gray conglomerates, sandstones, and siltstones are limited to the upper Norian-Rhaetian stage. These facts become clear under the assumption that during the Triassic period, this area was topographically elevated relative to the Karaburun and Kocaeli regions. It was a highland from where detrital material became transported as far as Izmir and Istanbul. In the uppermost Triassic it was flooded, except for several islands which existed as late as the lower Jurassic. This highland areally coincides with the western part of the North Anatolian welt (p. 32) which uplifted once more during Triassic time.

The Triassic in Middle Anatolia (Fig. 15)

Judged by the few localities, the Triassic system is quite uniform in this part of Turkey. At Ankara (*Chaput,* 1936: 241; *Bilgütay,* 1960a: 47) as well as farther east between Malatya and Erzincan, it is composed of gray to light-colored, bedded limestone and dolomite (*Kraus,* 1957: 39; unpub. data). South of Ankara a different facies is developed as limestones, red radiolarites, and pillow lavas (*Chaput,* 1936: 242).

According to its fauna and calcareous algae flora, all these deposits belong to the Ladinian and Late Triassic. The older series of the Triassic system have not yet been found. It appears, that major parts of Middle Anatolia have been subjected to a Ladinian (*Leuchs,* 1939) or Late Triassic transgression.

The Triassic in South Anatolia (Fig. 15)

Western Taurus

The Triassic system starts similar to Kocaeli with red, basal conglomeratic sandstones of probably Scythian age. In the west, at Milas and Bodrum, they emerge from the Permian which has a similar facies (*Brinkmann,* 1967: 5). In the northeast, in the foreland of the Sultan Dağ, they are overlying Paleozoic strata unconformably. A major transgression took place during the Anisian stage which created a geosyncline; its outline coincided roughly with that of today's western Taurus Mountains. The marginal areas of the geosyncline are characterized by gray dolomite and limestone, or sandy marl with remains of land plants and reef limestone lenses (*Brunn* et al., 1972a: 526). Toward the center of the geosyncline, the thickness of the Triassic strata decreases from 1000–2000 m to 50–100 m. Sediments of an open and gradually deepening sea were deposited in concentric zones: coral limestone and reef breccia (*Poisson,* 1967), reddish siliceous cephalopod-limestone (*Collignon* et al., 1970), and finally thin-bedded Halobia-radiolarian limestone, multicolored radiolarite, and turbidite (*Monod* et al., 1974: 118). The last three rock types are alternating in the central part of the geosyncline – in the *Pamphylian basin* (*Dumont* et al., 1972) – with basaltic lavas, in part pillow lavas, and tuffs. They are closely associated with ultramafitite (*Juteau,* 1969, 1970). The age of the sequence is Carnian, and possibly Norian (*Marcoux,* 1974; *Allasinaz* et al., 1974).

Eastern Taurus and Antitaurus

Scattered deposits, such as at Karaman (*Tuscu,* 1972: 14), at Alanya (*Blümel,* 1969; *De Peyronnet,* 1971: 95; *Argyriadis,* 1974), in the Bolkar Dağ (*Blumenthal,* 1955: 46), near Sariz in the Antitaurus (unpub. data), at Malatya (*Stchépinsky,* 1943: 117), and at Tarsus (unpub. data) show that in this area, too, the Triassic system is widespread. It is mostly composed of gray limestone and dolomite, and occasionally of reddish Daonella limestone.

The Triassic in Southeast Anatolia (Fig. 15)

The Scythian rocks are overlying the Permian conformably but with a hiatus (*Canuti* et al., 1970: 33) and consist of multicolored siltstone with sandstone, gray sandy marl and bedded limestone with a fauna poor in species. Towards the northwest, the thickness is reduced, and the clastic component is increasing. In the Middle Triassic, gray limestones with occasional sandy beds appear from which emerges a thick sequence of dark, bituminous limestone and dolomite, nearly devoid of fossils, that reaches as far as the Cretaceous (*Ten Dam,* 1955: 137; *Tolun,* 1962: 224; *G. C. Schmidt,* 1964: 109; *Altınlı,* 1966: 57; *Janetzko,* 1972: 9; *Lahner,* 1972: 74). Judged by the sediments, Southeast Anatolia was covered by a shallow, brackish bay during the Early Triassic epoch which opened towards the southeast. From the Middle Triassic onward, it became part of an euxinic ocean basin.

Since the ophiolite strata in northern Syria have yielded Late Triassic fossils (*Lapierre* and *Parrot,* 1972), one can assume that also the adjoining *Southeast Anatolian ophiolite zone* existed, entirely or in part, during the Triassic period.

Summary (Fig. 15)

The beginning of the Mesozoic era in Turkey and in the remaining Mediterranean region is characterized by a stratigraphic break. Lowermost Scythian marine deposits are unknown but appear farther east in Armenia and the Iran. The Triassic transgression of Turkey must have started from there.

The sea first advanced during early Scythian time across the central part of Syria and the Levantine Sea to the Aegean region and West Anatolia. In this period, parts of Southern Anatolia, in the Antitaurus (*Özgül* et al., 1973: 93), at Alanya (*Gedik,* pers. com.) and at Antalya (*Marcoux,* 1973), were flooded. About the same time, a shallow, brackish bay formed in northern Syria (*Wolfart,* 1967a: 35) and in Southeast Anatolia. Later, toward the end of the Scythian stage, the sea invaded from Bulgaria into Northwest Anatolia (*Paeckelmann,* 1938: 162).

Triassic

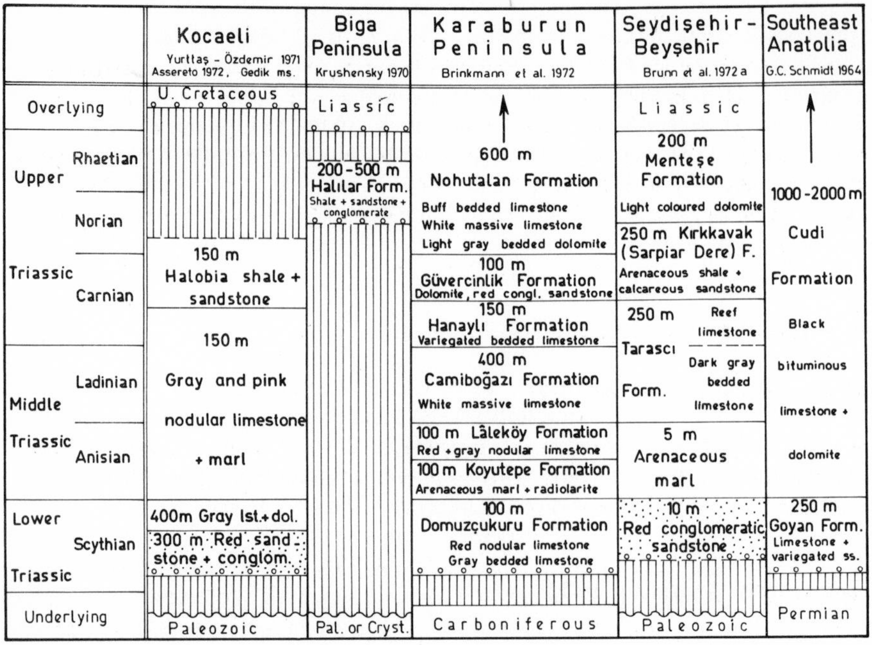

	Kocaeli (Yurttaş - Özdemir 1971, Assereto 1972, Gedik ms.)	Biga Peninsula (Krushensky 1970)	Karaburun Peninsula (Brinkmann et al. 1972)	Seydişehir-Beyşehir (Brunn et al. 1972 a)	Southeast Anatolia (G.C. Schmidt 1964)
Overlying	U. Cretaceous	Liassic	↑	Liassic	↑
Upper Triassic – Rhaetian		200-500 m Halılar Form. Shale + sandstone + conglomerate	600 m Nohutalan Formation. Buff bedded limestone, White massive limestone, Light gray bedded dolomite	200 m Menteşe Formation. Light coloured dolomite	1000-2000 m Cudi Formation. Black bituminous limestone + dolomite
Upper Triassic – Norian				250 m Kırkkavak (Sarpiar Dere) F. Arenaceous shale + calcareous sandstone	
Upper Triassic – Carnian	150 m Halobia shale + sandstone		100 m Güvercinlik Formation. Dolomite, red congl. sandstone; 150 m Hanaylı Formation. Variegated bedded limestone	250 m Tarascı Form. Reef limestone; Dark gray bedded limestone	
Middle Triassic – Ladinian	150 m Gray and pink nodular limestone + marl		400 m Camiboğazı Formation. White massive limestone		
Middle Triassic – Anisian			100 m Lâleköy Formation. Red + gray nodular limestone; 100 m Koyutepe Formation. Arenaceous marl + radiolarite	5 m Arenaceous marl	
Lower Triassic – Scythian	400m Gray lst.+dol.; 300 m Red sandstone + conglom.		100 m Domuzçukuru Formation. Red nodular limestone, Gray bedded limestone	10 m Red conglomeratic sandstone	250 m Goyan Form. Limestone + variegated ss.
Underlying	Paleozoic	Pal. or Cryst.	Carboniferous	Paleozoic	Permian

During the Anisian and the Ladinian stage, the three Early Triassic basins in West, Southeast, and Northwest Anatolia united into a single water body. At the end of the Triassic epoch, almost all of Turkey was below sea level. The Biga-Bursa rise had subsided; there are no signs of islands in the area of the Menderes and Kirşehir Massif. Only parts of Thracia and Northeast Anatolia may have been dry land.

The marine fauna reflects the progress of the Triassic transgression. It is Mediterranean in character with close relations to the Eastern Alps as well as to the Himalayas. Kocaeli shows some endemic species and a slight resemblance to Central Europe (*Assereto,* 1972: 441; *Jacobshagen,* 1972: 448).

During the Triassic period, the Alpidic geosynclines were formed on the broad floor of the Tethys which covered the eastern Mediterranean and Near East area. During the Triassic transgression, the *Pontian geosyncline* progressed from Bulgaria, along the Anatolian coast of the Black Sea, to the east, at least until Inebolu. The *geosyncline in the western Taurus* gained its shape; its central part turned into a deep trench with strong submarine volcanicity – an ophiolite trough (p. 84). It probably continued across Cyprus and Northwest Syria into neighboring Southeast Anatolia. A similar trench must have existed at Ankara, judged by the facies of the Triassic. It thus appears that several Anatolian ophiolite troughs date back to almost the beginning of the Mesozoic era. This may even be true for the majority of the troughs; observations by *Blumenthal* (1945) about pre-Liassic ophiolite sequences again deserve consideration.

Chapter 13 Jurassic

Jurassic fossils were first found by *De Tchihatcheff* (1848) at Ankara; *Pompeckj* investigated the Anatolian Jurassic system in 1897. The extent of the system in Turkey was not known until the travels of *Philippson* (1910–1915) and *Blumenthal* (1941–53). Paleontological studies were predominantly conducted on ammonites (lit. in *Bremer,* 1964, 1965, 1966), less on foraminifera (*Brönnimann* et al., 1970), sponges (*Gugenberger,* 1928), insects (*Laurentiaux,* 1946), and sporomorphs (*Ağralı* et al., 1965). Summaries have been compiled by *Chaput* (1936: 243) and *Bremer* (1964).

The Jurassic in Northeast Anatolia and Thracia (Fig. 16, 17)

The Liassic and Dogger in Northeast Anatolia

The two lower series of the Jurassic system show thicknesses of 1000 to 2000 m in a 150 km-wide zone that stretches from Inebolu across Amasya, Çorum, and Bayburt beyond Artvin along the valleys of the Kelkit and Coruh. They are mainly composed of clastic and volcanic rocks and rest on a base of old crystalline schists (*Blumenthal,* 1950: 35; *Ketin,* 1951: 118; *Fratschner,* 1954:214; *Altınlı,* 1956: 17; *Wedding,* 1963: 33; *Nebert,* 1964b: 42). The sedimentary part consists mostly of dark, grayish-green to black graywacke, sandstone, fine-sandy clay and marl; often the sediments become flysch-like. Current markings are frequent, whereas gradation is scarce (*Boccaletti* et al., 1968: 672). Basal conglomerates with granite, crystalline schist or marble pebbles are abundant, but conglomerate layers are also present in higher horizons. The volcanic part consists of basalt, andesite, and keratophyre in the form of sheets, veins, and sills. In addition, there are beds of finer and coarser tuff and tuffite. *Nebert* (1963: 7) ascribes all these rocks to submarine eruptions and thus explains their spilitization and green color.

This sequence contains various intercalations. First, there are red nodular marls with a rich ammonite fauna (p. 52). They appear repeatedly in several horizons of the Liassic and the lowermost Dogger. They are missing in the upper Dogger, here marine calcareous sandstones are present. Second, there are coal beds which, on the basis of sporomorph flora, are mostly of Liassic and occasionally of Dogger age (*Ağralı* et al., 1965). They are prevalent in the area between Kelkit, Bayburt, and Ispir. Third, red and gray siltstone with Estheria, insects, and land plants is known between Cide and Amasra (*Charles,* 1931: 161; *Laurentiaux,* 1946).

According to these characteristics, the Northeast Anatolian Liassic and Dogger rocks form a eugeosynclinal sequence. It fills an elongated basin that extends into the Black Sea west of Inebolu, and is termed *Inner Pontian geosyncline* (= North Anatolian Lias-Dogger trough; *Brinkmann,* 1968: 114). Most of the rocks belong to the Liassic series, which is particularly true for the lavas and tuffs. The Dogger ist less thick and contains only few tuff beds.

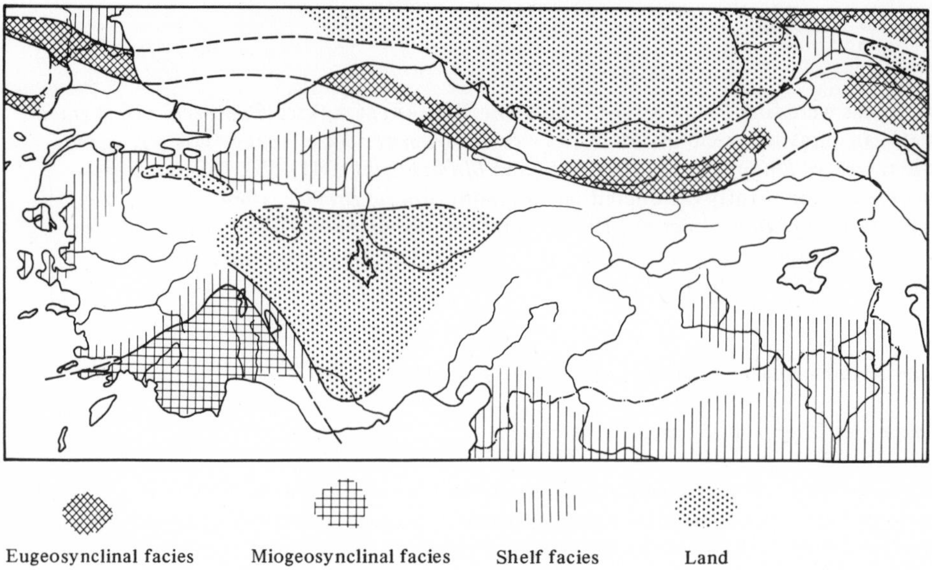

Eugeosynclinal facies | Miogeosynclinal facies | Shelf facies | Land

Fig. 16 Distribution, facies and paleogeography of the Liassic and Dogger in Turkey (easternmost part after *Vinogradov* et al., 1961)

The Malm in Northeast Anatolia

The Malm is unconformably – or at least with a hiatus – overlying the lower Jurassic rocks. At Küre, south of Inebolu, the Malm is transgressing with red, loose conglomerates which contain diabase and diorite pebbles. The latter came probably from the small diorite pluton of Küre which intruded into Liassic strata (*Kovenko,* 1944: 202; *Fratschner,* 1954: 220; *Boccaletti* et al., 1968: 672). Unconformities have also been described in the Amasya region (*Alp,* 1972: 39) and the upper Kelkit valley (*Ketin,* 1951: 119, 1959b: 84; *Baykal,* 1952: 296), whereas *Nebert* (1961a: 15) is explaining the relations by tectonic disharmony.

The following observations have been made on the time relationship of the angular unconformities:

Küre (south of Inebolu)
(*Grancy,* 1939; *Kovenko,* 1944)

Malm (including Kimmeridgian)
Lower Jurassic (including Sinemurian)

Kelkit-Siran
(*Baykal,* 1952)

Upper Malm/Lower Cretaceous
Lower Malm (including Lusitanian)
Lower Jurassic (including Liassic)

Bayburt
(*Ketin,* 1951)

Malm (including Oxfordian)
Lower Jurassic (including Sinemurian to Aalenian)

South of Kelkit
(*Kraus,* 1957)

Malm
Basal conglomerates (including Callovian)
Lower Jurassic

Amasra-Cide
(*Fratschner,* 1954)

Upper Malm/Lower Cretaceous
Dogger
Liassic

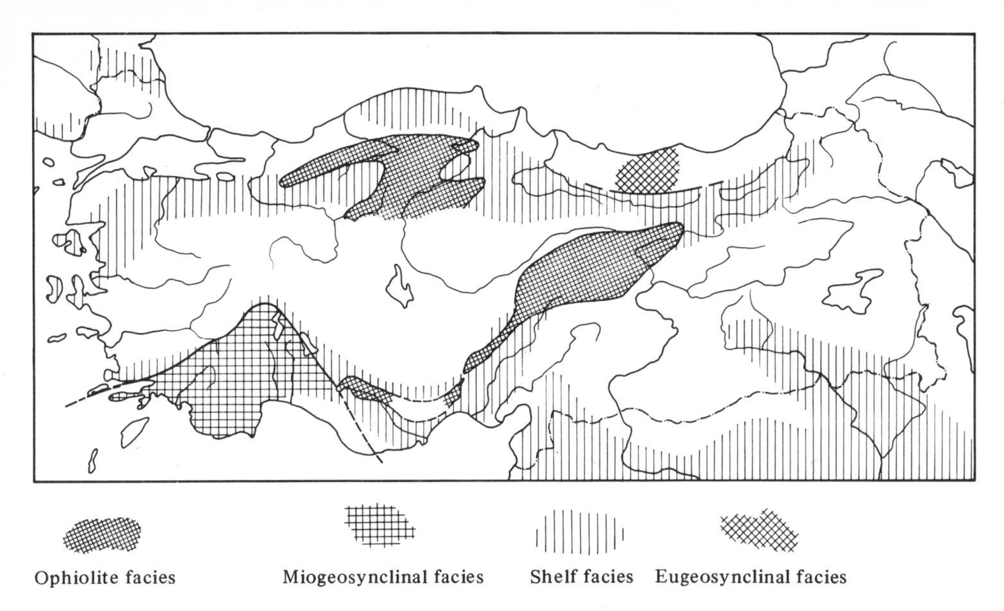

Fig. 17 Distribution, facies and paleogeography of the Lower Malm in Turkey

According to these observations, the middle Kimmerian as well as the late Kimmerian phase must have been active. It is probable that the Variscan granite of Gümüşhane (p. 38) has registered this tectogenesis because its biotite shows an age of 162 m. y. The diorite of Küre probably rose subsequent to the folding event. At the Kelkit river, too, granodiorite and quartz diorite have metamorphosed Liassic coal seams at their contact. They may also be of Kimmerian age, however, *Nebert* (1964b: 49) places them into the Late Cretaceaous epoch.

The prevalent Malm rocks consist of white to light gray, thickbedded to massive limestone. South of Giresun, the Malm contains basalt sheets (*Maucher* et al., 1962: 30, 65). It is conceivable that such intercalations are widespread in the Northeast Anatolian mountains. To the east, the limestones are progressively replaced by calcareous sandstones and quartz conglomerates (*Ketin,* 1951: 119; *Baykal,* 1951: 329; *Altınlı,* 1966: 43). The origin of the clastic material is unknown.

The Jurassic in Thracia

In the Istranca Massif, the old gneiss is distinctly overlain by rocks of the greenschist facies: phyllite, calcareous phyllite, and uppermost, light-colored marble (p. 5). Like in neighboring southern Bulgaria (*Bojadjiev,* 1971: 505), they represent metamorphosed Mesozoic strata which are very similar to the Jurassic rocks of the Inner Pontian Geosyncline. However, neither in Thracia nor in Bulgaria one finds angular unconformities below the Malm.

The Jurassic in Middle Anatolia (Fig. 16, 17)

Shelf Regions

From the northern to the middle part of Anatolia, the Liassic and Dogger are decreasing in thickness and in many places are totally missing. The volcanic component disappears except for some tuff beds. Lithologically, the lower Jurassic at Ankara is the same as in Northeast Anatolia; it consists of dark gray siltstones and sandstones with lenses of red nodular marl or crinoid-brachiopod-limestone (*Bremer,* 1965, 1966). Farther west, at Bilecik on the Biga Peninsula, the Liassic becomes coarser-grained and thicker, and in the upper Dogger, sandy iron oolite appears (*Altınlı,* 1963: 186; *Eroskay,* 1965: 144; *Altınlı* et al., 1970b: 79, 1972: 6). It is possible that the detritus of an island in the Ulu Dağ region has been deposited in this area.

The Malm covers a much larger space than the lower series of the Jurassic system and shows a development which is similar to that of Northeast Anatolia. In places where it emerges from the Dogger, its lowermost strata consist of sandy marl and yellowish-reddish limestone. In places where the Malm is overlapping, it begins with conglomeratic beds which contain fossils of the Oxfordian and Callovian stages (*Brinkmann,* 1971c: 60).

The bulk of the Malm consists of light yellow to light gray thick-bedded limestone. Foraminifera and calcareous algae occur frequently, whereas macrofossils, such as corals, brachiopods, and ammonites are scarce. The water depth increased during the Malm. Near Ankara where the entire Malm is composed of siliceous limestone and Aptychus limestone (*Bilgütay,* 1960a: 48), this happened at a very early stage. Elsewhere the Jurassic-Cretaceous boundary strata are frequently well-bedded Calpionella limestone (*Erk,* 1942: 73; *Akkuş,* 1971: 8).

There are three reasons why, different from the Malm, Liassic and Dogger are rare. First, the lower Jurassic strata show inherent gaps in their sedimentation. Characteristic fossils of the Hettangian stage have never been found in Turkey, and those of the Bajocian are only seldomly encountered (*Bremer,* 1965: 193, 1966). Resedimentation horizons with rounded fossils are mentioned occasionally. Second, before the deposition of the upper Dogger and Malm, an areal denudation of the lower Jurassic series had taken place. Bathonian, Callovian, and Oxfordian sediments are transgressive over vast areas of pre-Jurassic rocks. The uplift took place at approximately the same time as the Northeast Anatolian tectogenesis. Third, because of the mostly clastic development of the Liassic and Dogger, it is probable that during that period, the sea was restricted by land masses in Middle Anatolia. On the Biga Peninsula and at Bursa (p. 42), the Late Triassic archipelago may have continued to the Dogger. A larger island was situated south of Ankara. Here, Dogger and Malm are advancing to the south beyond the Liassic sediments. Arkose with thin coal seams and detrital crinoidal limestone indicate the proximity of a coastal area (*Bremer,* 1965: 196, 1966: 156).

Most of this area has subsided below the level of the sea during the Malm. Middle Anatolia was nearly or entirely covered by an open sea whose depth increased towards the close of the Jurassic period.

Ophiolite Troughs

The Middle Anatolian shelf sea of the Jurassic period was dissected by several deep troughs in which the ophiolite sequence, known since Triassic time, was deposited. Such a trough extended from Mudurnu southwest of Bolu, where chert and tuffite with small serpentinite and spilite pebbles are underlying Tithonian limestone (*Abdüsselâmoğlu,* 1959: 31; *Boccaletti* et al., 1966b), to the area of Cankırı and Alaca northeast of Ankara, where the ophiolite strata contain intercalations of Aptychus and Calpionella limestone (*Nowack,* 1928: 306; *Boccaletti* et al., 1966a: 496; *Bortolotti* and *Sagri,* 1968: 663). A second trough was probably running from Erzincan, along the western slope of the Antitaurus, toward Tarsus. In the Munzur Dağ, neritic limestone of the upper Malm is overlying ultramafic and ophiolitic rocks (*Olson,* pers. com.; unpub. data). The same holds true for the area of Kangal and Pınarbası between Kayseri and Sivas (unpub. data), as well as for the area of Namrun near Tarsus (unpub. data). At Hekimhan north of Malatya (*Izdar,* 1963: 4), in the western Antitaurus (*Vohryzka,* 1966: 99), at the Bolkar Dağ (*Blumenthal,* 1955: 60), and at Pozanti (*Ovalioğlu,* 1963: 3) the ultramafic bodies are covered by littoral-neritic Upper Cretaceous sediments. But here, too, the formation of the ophiolite sequences may have come to an end at an earlier date. Thus, folded radiolarite of possibly Jurassic age is preserved between the transgressive Campanian sediments and the ultramafitic base at Hekimhan.

The Jurassic in West and South Anatolia (Fig. 16, 17)

West Anatolian Coast

On Karaburun and on Chios, limestone sedimentation is prevalent during the entire Jurassic period; however, the thick sequence, poor in fossils, can only be roughly subdivided. Liassic and Malm deposits have been found. It is possible that part of the Dogger is missing, and that this gap is marked by a narrow bauxite horizon (*Brinkmann* et al., 1972: 145; *E. Flügel,* 1974).

Western Taurus

Sedimentation generally continues from the Triassic into the Liassic. Only west of Antalya *Kalafatcioğlu* (1973: 78) describes an unconformable overlap of the Jurassic beds.

The bay which since Triassic time occupied the place of the present-day mountain range, is also prominent in the facies distribution of the Jurassic, and in particular of the Liassic. At the very western boundary, at Bodrum, and on the Marmaris Peninsula, as well as at the very eastern boundary, at Anamur, a gap exists in the lower Jurassic strata (*Orombelli* et al., 1967: 832; *Brinkmann,* 1967: 7; *Isgüden,* 1971: 41). Complete Jurassic profiles can only be found as far as Fethiye in the west and Alanya in the east. At the northwestern edge of the bay, the Liassic is represented by black, sandy shale. At the northeastern edge, at the Beysehir Gölü and at Akseki, the lower Liassic rocks consist of red, conglomeratic sandstone and gray, marly limestone (*Brunn* et al., 1972a: 527; *Desprairies* and *Gutnic,* 1972: 506). During the Dogger, and particularly the Malm, the basin expanded. But even in the Kimmeridgian stage, the mainland was not far away in the northeast. During a period of emergence, basalt sheets which are intercalated in marine strata southwest of the Sultan Dağ, weathered into bauxite (*Desprairies* and *Gutnic,* 1972: 511; *Gutnic* and *Juteau,* 1973). At Akseki, fissile, marly limestones contain ammonites as well as remains of gymnosperms (*Corsin* and *Martin,* 1969; *Enay* et al., 1971).

In the inner parts of the bay, the thicknesses of the Jurassic as well as the Triassic strata are decreasing to a few tens of meters (*Gutnic* and *Monod,* 1970). Clastics are missing. In the Liassic, red nodular and fine-grained limestone with cephalopods, in the Dogger, oolitic limestone with thick lenses of white coral limestone, and in the Malm, light-colored, bedded limestone with chert nodules are prevalent (*Poisson,* 1968, 1974; *Lefèvre* and *Marcoux,* 1970; *Brönnimann* et al., 1970; *Brunn* et al., 1972a: 530; *Baykal* and *Kalafatçioğlu,* 1973: 37). Jurassic radiolarites are known in the surroundings of Antalya; but the typical ophiolitic association is missing (*Dumont* et al., 1972). Thus, during the entire period, the inner part of the Taurus geosyncline was covered by an open sea which was divided into rises and basins; the Pamphylian ophiolite trough (p. 43), however, did not exist in the Jurassic period.

Eastern Taurus

Apparently Jurassic rocks are missing along the southern shore between Silifke and Mersin as well as in the Bolkar Dağ and Aladağ. At Karaman, Liassic and Malm strata are known in a limestone sequence which extends from the Triassic to the Cretaceous (*Tuscu,* 1972: 16).

The Jurassic in Southeast Anatolia (Fig. 16, 17)

The Jurassic as well as the Triassic system are represented by the Cudi formation; gray to black, bituminous, bedded to massive limestones and dolomites with intercalations of marly and siliceous rocks (*Tolun,* 1951: 73, 1962: 225; *Altınlı,* 1966: 57). It is quite dominant along the southern border of Turkey, at Hakkâri, Siirt, and in the Amanos Dağ (*S. Türkünal,* 1953: 10; *Janetzko,* 1972: 9; *Lahner,* 1972: 74). Toward the north, the formation is decreasing in thickness and time coverage. At Hazro, it is only 100 m thick; and at its base, *Tolun* (1949: 76) collected middle Liassic fossils. The Southeast Anatolian ophiolite trough probably existed throughout the Jurassic period, but its documentation is uncertain (*Rigo De Righi* and *Cortesini,* 1964: 1923).

Summary (Fig. 16, 17)

In Turkey, the Jurassic and Triassic systems are generally separated by a regression. At the beginning of the Liassic stage, the major part of the country was dry. Only the West Anatolian coast, the geosyncline of the western Taurus, and Southeast Anatolia remained under water beyond the Triassic-Jurassic boundary. But even here, short-lived regressions occurred, as shown by the profiles near Beysehir and Hazro.

During the Jurassic period, two transgressions took place which were interrupted by an interval of receding shorelines. The Liassic transgression flooded only parts of Turkey; in Thracia and Central Anatolia land masses remained. The outline of a Central Anatolian island, comprising the present-day Kırşehir Massif and the subsurface of the Tuz Gölü Basin, can be traced by the Liassic littoral deposits near Ankara and at the Sultan Dağ, respectively. It is probable that there were other, smaller islands, as, for instance, in the region of the Ulu Dağ. The Liassic transgression reached its peak during the Domerian-Toarcian stage. Then the sea receded, especially during the Bajocian stage. The second Jurassic transgression started in the Bathonian and lasted until the Kimmeridgian-Portlandian. Judged by the generally uniform facies of the Malm, the major part of Turkey must have subsided below sea level. But it is possible that islands with a low morphological relief might have existed in the place of the Menderes, Kırşehir, and Bitlis Massifs.

From the Triassic to the Jurassic period, the epirogenic division of the Tethys floor became more accentuated. The older geosynclines became more pronounced, and new geosynclines developed. Going from north to south, the following geotectonic units can be distinguished.

1. During the Early Jurassic epoch, the Pontian land must have risen rapidly, because in the adjacent geosynclinal areas to the north and south collected vast amounts of clastic material. Its thickness reached more than 1000 m in the Inner Pontian geosyncline and many times more than that in the Dobruja-Crimea-Caucasus geosyncline (*Brinkmann,* 1974: 67). The

Jurassic

	Biga Peninsula Brinkmann 1971 c	Karaburun Peninsula Brinkmann et al.1972	Ankara Bremer 1965, 1966	Upper Kelkit Valley Nebert 1961 a, 1964 b Wedding 1963	Pisidian Taurus Martin 1969	Southeast Anatolia Tolun 1962 G.C. Schmidt 1964
Overlying	Neogene	Lower Cretaceous				↑
Malm	300 m White thick bedded limestone — 1 m Reddish limestone	200 m Gray Cladocoropsis limestone	300 m Light coloured bedded limestone — 10 m Light gray nodular limestone	300 m White thickbedded limestone	350 m Akköy Form. White cherty limestone / Thin bedded marl — 400 m Hendos Formation Light coloured thick bedded dolomitic limestone	1000–2000 m Cudi (Tanin) Formation Dark gray bituminous limestone + dolomite
Dogger	50 - 100 m Brown arenaceous lms.+ iron oolithic ss	250 m Upper Nohutalan Formation Buff limestone	50 m Greenish gray sandstone + shale — 10 m Red nodular marl	300 m Gray calcareous sandstone + variegated siltstone	250 m Yellow arenaceous limestone	
Lias	50 m Gray siltstone		40 m Gray siltstone — 15 m Red nodular marl — 30 m Calc. sandstone	1000 - 1500 m Black arenaceous shale with volcanics and coal seams	300 m Üzümdere Formation Variegated conglomeratic sandstone + marl	↑
Underlying	Paleozoic or Cryst.	Upper Triassic	Pal. or Triassic	Cryst. schist	?	

Pontian land was almost restricted to the central part of the present-day Black Sea, but temporarily it extended to the south into the area of the Pontian geosyncline. The paleogeographic relations of the Lower Jurassic rocks are in many instances comparable to those of the Upper Carboniferous rocks which is particularly true for the coal beds.

2. In the Lower Jurassic, the Inner Pontian geosyncline had been fully developed as a eugeosyncline. It crossed Northeast Anatolia and extended into the Black Sea west of Inebolu. In the Istranca Massif, it probably emerged again and continued into southern Bulgaria. The geosyncline was short-lived; in Northeast Anatolia it ended with the middle or late Kimmerian phase. The formation of the *Inner Pontides* was accompanied by the rise of several granitic-dioritic plutons.

3. The middle part of Anatolia, on the whole, represented geotectonically a geanticline, and paleogeographically a shallow sea with some islands. But the sea floor was dissected by several, deep ophiolite troughs. Information about their areal extent and age is still incomplete.

4. Since the Triassic period, the geosyncline of the western Taurus existed. The predominantly calcareous sedimentation continued throughout the Jurassic period. There is no indication of an ophiolite trough in the central part of the geosyncline. The volcanic activity was limited to several basalt eruptions at the Sultan Dağ and near Antalya.

5. Major parts of Southeast Anatolia belonged to a semi-euxinic sea, whose sediments extend into the Iraq and northern Syria. No conclusive evidence exists concerning the Southeast Anatolian ophiolite trough during the Jurassic period.

During the Jurassic, and particularly during the Malm, Turkey was covered by a sea which extended far beyond the country's boundaries. To the north, a land mass was situated in the Black Sea. To the south, all of the Near East, major parts of Arabia, and the Sinai Peninsula were submerged. These oceanic connections explain the close lithological and paleontological relations of the Turkish Jurassic system with its neighboring areas. Already *Pompeckj* (1897) pointed out the similarity between the sandy older Jurassic rocks of Anatolia and the Gresten strata of the Eastern Alps and the Balkan Mountains. Red limestones in the Ammonitico Rosso, Adnet and Hierlatz facies are widespread in the Jurassic system of Turkey. The Malm limestone resembles the Quinten limestone in the Swiss Alps. The ammonites of northern Anatolia belong to the Mediterranean-Caucasian province (*Pompeckj,* 1897: 754; *Pia,* 1913: 386; *Gugenberger,* 1928: 279). The ammonite fauna of southern Anatolia, on the other hand, is most closely related to that of the Himalaya and the Gondwana regions (*Enay,* 1975).

Chapter 14 Cretaceous

At Izmir in 1840, *Strickland* collected the first rudists and thus proved the existence of the Cretaceous system in Turkey. Its vast extent, particularly in the north and east, was first recognized by *De Tchihatcheff* (1866–69). Descriptions of parts of the Cretaceous marine fauna have been given by *Broili* (1911), *Frech* (1910, 1916), *Böhm* (1927), *Astre* and *Charles* (1931), *Lambert* (1931), *Nöth* (1931), *Kühn* (1933), *Lambert* and *Charles* (1937), *Delpey* (1938), *Stchépinsky* (1942), *Tilev* (1951), *Pınar* (1954, 1956), *M. Türkünal* (1958), *Öztemür* (1959), *Meriç* (1967), *Durand-Delga* and *Gutnic* (1966), *Lefèvre* and *Sornay* (1967), and *Karacabey* (1972). For summaries of the geological literature see *Chaput* (1936: 245) and *Altınlı* et al. (1969).

The Cretaceous in Thracia (Fig. 18)

Cretaceous strata have been found at two localities. At the coast of the Black Sea close to the Bulgarian border, near Igneada, rudist limestone and sandy marl are overlying the crystalline rocks of the Istranca Massif, starting with a basal conglomerate of Cenomanian age (*Pamir* and *Baykal,* 1947: 32). They belong to the southern marginal area of the Bulgarian Srednogorie zone. In the Tekirdağ, *Kopp* et al. (1969: 49) found red and greenish Globotruncana marl of the Campanian-Maestrichtian stage. There may be a connection with similar deposits in Northwest Anatolia.

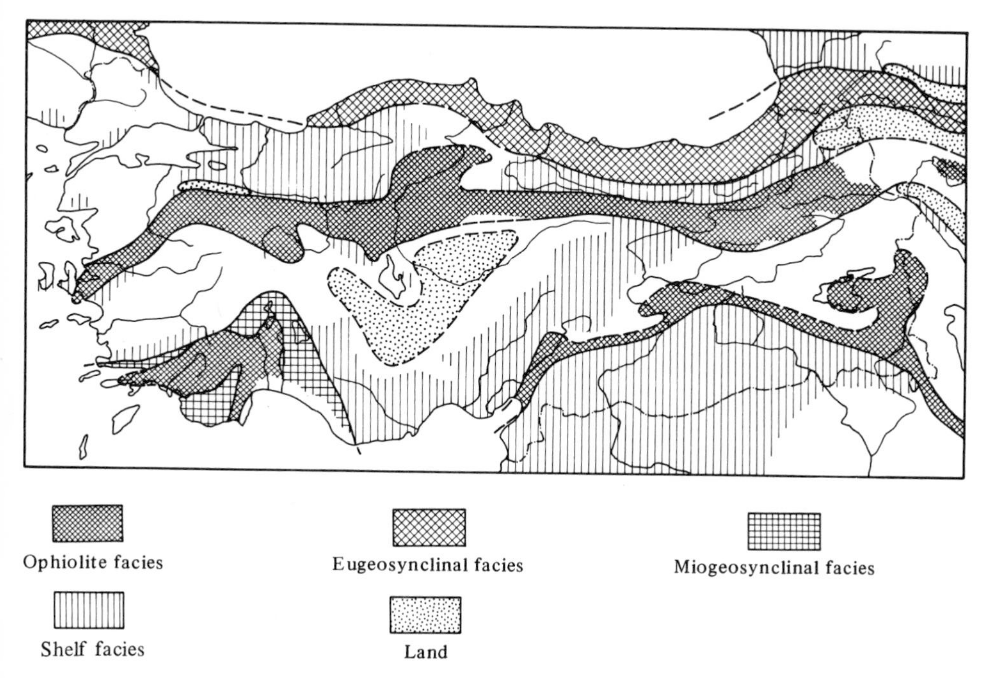

Fig. 18 Distribution, facies and paleogeography of the middle Cretaceous in Turkey (easternmost part after *Vinogradov* et al., 1961)

The Cretaceous in North Anatolia (Fig. 18)

In this area, the Cretaceous cover of the bituminous coal deposits of Zonguldak (*Arni,* 1931; *Tokay,* 1952: 45; *Altınlı,* 1951c: 307) and Amasra (*Charles,* 1930; *Altınlı,* 1951a: 159, 1951c: 307; *Baykal,* 1954: 193; *Tokay,* 1954/55: 51; *Ketin,* 1955a: 148) is best known. At Zonguldak, a coarse-grained conglomerate, consisting mainly of Lower Carboniferous limestone detritus, is covering the Paleozoic rocks in angular unconformity. It is overlain by massive limestone layers in Urgonian facies, whose Barremian-Aptian fauna suggests the time of transgression. The middle Cretaceous rocks are composed of sandstone and marl. During the Cenomanian stage, they emerged into a sandstone flysch which changed into a wildflysch with huge boulders of Devonian and Lower Carboniferous sedimentary rocks, Upper Carboniferous coal, as well as Lower Cretaceous limestone. It is likely that landslides spread out on the sea floor. This process may have been initated by tectonic movements of the Austrian phase; the Turonian rocks are often overlapping unconformably. The younger Cretaceous series consists of a thick limestone flysch: white, gray, and pink to red, graded calcarenites whose thin beds are separated by layers of marl. This sequence extends to the end of the Upper Cretaceous; only in the Campanian stage it is locally interrupted by an unconformity. During the Turonian stage, and at some places already during the Albian-Cenomanian stage, the first pyroclastic intercalations appear. During the Santonian stage, the volcanic activity reached its maximum with submarine eruptions of andesite and basalt as well as coarse- and fine-grained tuff. In the Campanian, the volcanic rocks are decreasing in volume, and they are rare in the Maestrichtian. The transition from the Cretaceous to the Tertiary system is without a hiatus, and the limestone flysch at first turns marly and finally sandy. *Dizer* (1972) and *Sirel* (1973) have classified the boundary strata micropaleontologically.

West of Ereğli, the Lower Cretaceous rocks are thinning out, and on Kocaeli only the younger Upper Cretaceous rocks are transgressive (p. 55). To the east, however, the profile becomes stratigraphically more complete. Jurassic deposits are present from Cide onward. South of Inebolu and Sinop, the transition from the Jurassic to the Cretaceous system takes place in a sequence of light-colored limestone (*Grancy,* 1939: 79; *Wedding,* 1970: 45). Malm has been found as far as Giresun-Gümüşhane. Farther to the east, it seems to vanish, and at Artvin, volcanic Cretaceous rocks are overlying the crystalline basement. Close to the Black Sea coast the Upper Cretaceous flysch is 2000 m thick, and decreases to 1000 m farther inland. Slumping and olistostromes are widespread. The current markings are pointing mainly to the northeast and east (*Boccaletti* et al., 1968: 672; *Sestini* and *Canuti,* 1968: 320). The volume of volcanic rocks is increasing from Samsun toward the east. Near Giresun, Trabzon, and Artvin, calcareous marine sediments only occur in the Campanian-Maestrichtian stage. The remainder of the sequence is composed of volcanites whereby tuffs and agglomerates are outnumbering sheets and sills. Although pillow lava seems to be missing, the alternation of volcanites and sediments argues for submarine eruptions. However, ignimbrite has also been found, according to *Kilinç* (1971: 54). Throughout the Cretaceous system, andesitic-basaltic and dacitic eruptives are alternating with each other (*Maucher* et al., 1962). Toward the south, the thickness of the Cretaceous strata is decreasing, in Late Cretaceous time the limestone flysch is replaced by sandstone flysch, and the volcanic intercalations are restricted to the Maestrichtian stage (upper Kelkit valley; *Nebert,* 1961a: 18).

This suggests that during the Cretaceous period a eugeosynclinal subsidence zone existed along the Anatolian coast of the Black Sea which is here named the *Outer Pontian geosyncline.* In Northeast Anatolia, volcanism started as early as during the Malm; toward the west it began at a later time; at Zonguldak it did not commence until the Upper Cretaceous. In general, the thickness of the strata and the volume of volcanic matter are increasing toward the present coast. It can therefore be deduced that between the Turkish eastern boundary and Ereğli, the geosyncline rested only partly on the present-day mainland. West of Ereğli, it disappears entirely below the Black Sea.

The Cretaceous in Middle Anatolia (Fig. 18)

Shelf Environments

On Karaburun, Lower Cretaceous coquina limestone is overlying Cladocoropsis limestone of the Malm. Both formations are conformable and separated by a thin layer of bauxite (*Brinkmann* et al., 1972: 141).

In general, however, the transition from the Jurassic to the Cretaceous system occurs without interruption. In many places, the facies of the Malm continues into the Lower Cretaceous, and the boundary between the two systems is in a sequence of light-colored bedded limestone (*Kalafatçıoğlu* and *Uysallı,* 1964: 3; *Nebert,* 1964b: 48; *Akkus,* 1971: 6). Compared to the Malm, the Cretaceous sediments are marly and show thinner bedding; ammonites become more frequent. The Berriasian, Valanginian, Hauterivian, and Barremian stages are widely present, and the Aptian and Albian stages are found at some places (*Blumenthal,* 1950: 48; *Ketin,* 1951: 119; *M. Türkünal,* 1959: 73; *Açar,* 1970: 8). Then follows a hiatus, the extent of which is increasing with the distance from the southern edge of the Pontian geosyncline. At the upper Kelkit river (*Nebert,* 1964b: 48), southwest of Bolu (*Abdüsselâmoğlu,* 1959: 40), at Adapazarı and Şile (*Baykal,* 1942: 184), Cenomanian or Turonian sediments are overlapping. The southern and western parts of Kocaeli were not submerged until the Santonian stage. Following a short period of regression, the widest advance of the Cretaceous sea occurred during the late Campanian-Maestrichtian stage. Compared with the uniform facies of the Lower Cretaceous series, the sediments of the younger Upper Cretaceous are quite varied. In southern Thracia and Northwest Anatolia, bathyal foraminiferal marl of the Couches Rouges type (*Brinkmann* et al., 1976) is present. Outcrops of conglomeratic rudist limestone and light-colored, chalky marlstone occur on the southern coast of Kocaeli (*Altınlı* et al., 1970a: 72). On the northern coast, closer to the Pontian geosyncline, dacitic tuffs strongly participate (*Chaput,* 1936: 152, 245). Between Bursa and Bilecik, the Campanian rocks are composed of graywacke and sandstone containing mica schist pebbles which, as in the Jurassic (p. 48), probably originated from an Ulu Dağ island. The Maestrichtian stage is partly represented by light-colored, bedded Orbitoides limestone and partly by brownish, fossil-rich calcareous sandstone (*Altınli* et al., 1970b: 80, 1972: 11). At Amasya, *Blumenthal* (1950: 53) found Hippurites-Actaeonella limestone and Cyclolites marl, similar to the Gosau formation in the Eastern Alps.

From the Cenomanian onward, a bay advanced from the north to Haymana south of Ankara and to the area of the Tuz Gölü. It represented the beginnings of the Tuz Gölü Basin (Turk. Gulf Oil Co., 1961). During the Campanian-Maestrichtian, flysch-like sandstone was deposited, which contains a rich marine fauna but also some thin coal seams. The clastic material came partly from a crystalline massif and partly from an ophiolite zone (*Chaput,* 1936: 55; *Yüksel,* 1970). The Middle Anatolian island (p. 50) thus still existed.

During the two regressive intervals, at the beginning of the Upper Cretaceous and in the Campanian, slight crustal movements occurred:

Mudurnu southeast of Adapazari
(*Abdüsselamoglu,* 1959)

Lower Eocene
~~~~~~~~
Paleocene
~~~~~~~~
Campanian-Maestrichtian to
Cenomanian-Turonian
~~~~~~~~
Lower Cretaceous
(Valanginian-Hauterivian)

Nallıhan west of Ankara
(*Kalafatçıoğlu* and *Uysallı,* 1964)

Paleocene
Maestrichtian
Campanian
? Turonian-Santonian
~~~~~~~~
Lower Cretaceous
(Valanginian to Barremian)

which probably belong to the Austrian and Ressen phases (*Erol,* 1961: 77). It appears that the magmatic processes were of more significance than the tectogenic ones. Radiometric dating of several west and central Anatolian granodiorites gave ages between 68 and 74 m. y. (*Çoğulu* et al., 1965; *Bürküt,* 1966; *Çoğulu* and *Krummenacher,* 1967; *Bingöl,* 1971). The question remains wether these ages indicate formation or reheating of the plutons.

Ophiolite Troughs

As in earlier Mesozoic time, the eugeosynclinal ophiolite facies was the equivalent of the neritic-bathyal development and had a particularly wide areal extension during the Cretaceous.

The *Middle Anatolian ophiolite zone* (*Brinkmann,* 1968: 116) dissects all of Anatolia along the line Izmir-Ankara-Erzurum. During the Cretaceous period, this structure progressed both to the east and to the west. The middle part of the zone, between Bursa and Yozgat, existed from the Triassic to the Late Cretaceous (*Arni,* 1942: 483; *Ketin,* 1955b: 30; *Erol,* 1956: 21; *Boccaletti* et al., 1966a: 494; *Bortolotti* and *Sagri,* 1968: 664; *Sestini,* 1971: 374; *Alp,* 1972: 65; *Norman,* 1972: 183, 240). The eastern part, between Sivas and Erzurum, has furnished only Early and Late Cretaceous fossils (*Ketin,* 1945: 291, 1951: 120; *Baykal,* 1947: 200, 1951: 329; *Blumenthal,* 1958: 253; *Altınlı,* 1966: 44; *Tatar,* 1971: 81). The western part, between Bursa and the Aegean coast, may even be younger. At Izmir, the ophiolite strata belong to the Maestrichtian stage (*Akartuna,* 1962a: 4; *Verdier,* 1963: 39).

These tentative age assignments are supported by the stratigraphic position of the beds overlying the ophiolite sequences. All of them correspond with the latest Cretaceous or the earliest Tertiary period:

Akhisar north of Izmir
(*Dizer,* 1962a; *Verdier,* 1963)

Lower Eocene
~~~~~~~~~~
Ophiolite stratum (including Maestrichtian)
Cenomanian-Maestrichtian limestone

Tavsanlı west of Kütahya
(*Kaya,* 1972)

Orbitoides limestone (Maestrichtian)
Sandstone flysch (Turonian-Campanian)
Ophiolite stratum

East and northeast of Ankara
(*Blumenthal,* 1948; *Boccaletti* et al., 1966a; *Norman,* 1972)

Eocene
Paleocene
Flysch (Maestrichtian)
Ophiolite stratum (including Malm-Campanian)

Rudist limestone (Maestrichtian)
~~~~~~~~~~
Ophiolite stratum (including Triassic, Jurassic, Lower Cretaceous-Turonian)

Eocene
~~~~~~~~~~
Ophiolite stratum (including Campanian-Maestrichtian)

Yozgat
(*Ketin,* 1955b)

Lower Eocene
~~~~~~~~~~
Paleocene sandstone
Ophiolite stratum (including Turonian-Maestrichtian)

Northwest of Sivas
(*Tatar,* 1971)

Upper Paleocene
~~~~~~~~~~
Sandstone flysch
Ophiolite stratum (including Cenomanian-Turonian)

Bayburt
(*Ketin,* 1951)

Rudist limestone (Campanian-Maestrichtian)
~~~~~~~~~~
Ophiolite stratum
Lower Cretaceous limestone (Valanginian-Albian)

According to this, the development of the eugeosynclinal trough of the Middle Anatolian ophiolite zone ended no later than Paleocene. In some places, this happened in such a way that the ophiolite strata gradually developed into shallow-water sediments. In most instances a tectogenesis taking place during the Ressen and occasionally during the Laramian phase terminated the ophiolite sequence. Following the tectogenesis, the ophiolite zone rose and became eroded. Mainly in the central and eastern segments of Middle Anatolia, uppermost Cretaceous sediments composed of detritus of the ophiolite series are frequent. They range from coarse-grained serpentinite-spilite-radiolarite conglomerates to sandstone to multicolored fine breccia and are closely associated with rudist limestone, Orbitoides limestone, and andesitic-basaltic lava (*Pamir* and *Baykal,* 1943: 313; *Ketin,* 1945: 292; *C. Erentöz,* 1954: 33; *Blumenthal,* 1955: 60; *Izdar,* 1963: 8; *Akkuş,* 1971: 10; *Aykulu,* 1971: 33; *Norman,* 1972, 1973a; *Leo* et al., 1974).

The connection of the ophiolite area between Ankara, Konya, and Karaman with the Middle Anatolian zone is obscured by a Tertiary cover. The rocks presumably belong mostly to the Upper Cretaceous (*Blumenthal,* 1956: 8, 14; *Wirtz,* 1958: 331; *Passerini* and *Sguazzoni,* 1966: 511; *Tusçu,* 1972: 18).

The Cretaceous in South Anatolia (Fig. 18)

Western Taurus

During the Cretaceous, the geosyncline showed a similar outline as in earlier Mesozoic time: the deposits are mainly calcareous, too. In the early and middle Cretaceous period, bedded limestone occasionally containing Chara remains in addition to marine calcareous algae, was deposited in the marginal areas (*Brunn* et al., 1972a: 529). At the northeastern edge of the basin, the coastline retreated during the earlier Upper Cretaceous epoch. In the dry area at Akseki, a bauxite deposit formed with a thickness of 1–2 m (*Martin,* 1969: 116). During the same period, other parts of the geosyncline became temporarily shallower; rudist limestone is widespread during the Santonian stage. Starting with Campanian time, the water depth increased again and the sea expanded. Yellowish-reddish marly Globotruncana limestone interfingering with Orbitoides limestone can be found even in the formerly marginal areas (*Brunn* et al., 1972a: 528 f.).

The geosyncline was divided into two basins which were separated by the submarine rise of the Bey Dağlari southwest of Antalya. The western, *Lycian Basin* between Denizli and Fethiye turned into an ophiolite trough in which a wildflysch with huge boulders was deposited temporarily. The sequence ended in the Paleocene or perhaps Eocene (*De Graciansky,* 1972: 239). The eastern, Pamphylian Basin may have been revived (p. 43) as an ophiolite trough during the Cretaceous. Southwest of Antalya, *Colin* (1962: 49) observed rocks of that series in connection with middle Cretaceous strata, whereas *Dumont* et al. (1972: 397) and *Marcoux* and *Poisson* (1972) mention only radiolarites but no magmatites in this area. Here, too, strong crustal movements started at the end of the Cretaceous period, as evidenced by intercalations of ophiolite detritus in Maestrichtian sediments.

Eastern Taurus

While at Akseki north of Alanya the Cretaceous system is still 1000 m thick and stratigraphically nearly complete, it decreases east of Anamur to 200 m of bedded limestone which represents only a part of the Cretaceous system (*Blümel,* 1969: 437; *Işgüden,* 1971: 44). On the southern coast at Silifke (*Yalcınlar,* 1973: Fig. 21) and in the Antitaurus (*Baykal,* 1945: 137), the situation is similar.

The Cretaceous in Southeast Anatolia (Fig. 19, 20)

The dark-colored limestones and dolomites of the Cudi Formation in almost unaltered facies extend from the Triassic (p. 43) to the middle Cretaceous. Then a reversal occurred. A geanticline, the *Mardin welt,* rose in the central part of Southeast Anatolia; in its core lower Paleozoic strata were exposed (*Rigo De Righi* and *Cortesini,* 1964: 1928). The uplift lasted until the Upper Cretaceous epoch, and, as a consequence, the sea remained particularly shallow above the welt following the renewed flooding during Albian-Cenomanian time. These conditions favored the formation of mineable phosphate deposits (*Beer,* 1967). In the course of the Upper Cretaceous, the water depth increased in the entire area (*Cordey,* 1971; *Keskin,* 1971b).

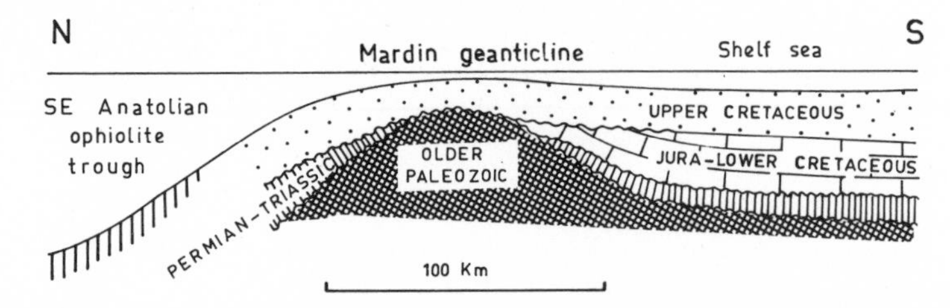

Fig. 19 Schematic section across Southeast Anatolia in the Upper Cretaceous period. Mardin geanticline still acting as a positive structure (after *Rigo De Righi* and *Cortesini,* 1964)

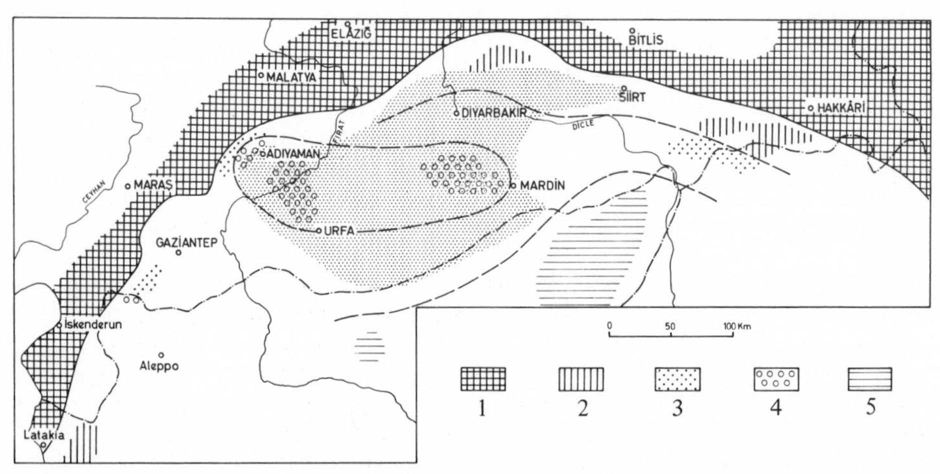

1 = Ophiolite, 2–5 = Facies of the middle Upper Cretaceous (2 = Siliceous limestone, 3 = Phosphatic glauconitic limestone, 4 = Phosphate, 5 = Marl)

Fig. 20 Facies distribution in the Cretaceous deposits of Southeast Anatolia. Shaded area = Mardin geanticline in the subsurface (after *Beer,* 1967)

North of the Mardin welt, the Southeast Anatolian ophiolite trough extended in an arc ranging from Iskenderun to Adıyaman and Hakkari. Although it already existed during earlier Mesozoic time, its main content was deposited during the Upper Cretaceous and Paleocene (*Ketin,* 1950: 155; *Dubertret,* 1955: 97; *Rigo De Righi* and *Cortesini,* 1964: 1923; *Hall* and *Mason,* 1972; *Roloff,* 1972: 103). At the same time, the trough reached its widest expansion; even the area of Malatya, Elaziğ, and Van was included (*Chaput,* 1936: 251, 1939;

Ketin, 1947a: 260; *Demirtaşlı* and *Pisoni,* 1965: 28). The Bitlis Massif, too, became flooded, but it remained a submarine rise (*Tolun,* 1951: 84). In the Campanian stage, tectogenic processes started. In the western part of the area, between Iskenderun and Adıyaman as well as at Malatya and Elaziğ, this process was achieved mainly in one act. Sandy Maestrichtian marl is transgressive with basal conglomerates. In the east, at Siirt and Van, the movements lasted until the earliest Tertiary epoch. Starting with the Maestrichtian stage, the ophiolite facies was replaced by deposits of clastic material, sandstone flysch and red conglomerates, derived from the ophiolite zone and the Bitlis crystalline Massif (*Rigo De Righi* and *Cortesini,* 1964: 1921).

Summary (Fig. 18)

In general, the Cretaceous strata of Turkey are conformable with the Jurassic and Tertiary rocks. The Cretaceous period, however, is divisible into three natural stages by two important tectogeneses – the Austrian phase at the turn of the Lower to the Upper Cretaceous, and the Ressen phase in the Campanian. The Lower Cretaceous represented a continuation of the Malm. The sediments and paleogeography were only slightly modified. At the end of the Early Cretaceous epoch, the sea receded, and bauxites appear at several places: at Karaburun, the western Taurus, near Mersin and Iskenderun (*Wippern,* 1962, 1964b). The Upper Cretaceous started with a transgression which lasted until the Santonian. It affected the northwestern, central, and southeastern parts of Anatolia, as well as parts of the Taurus geosyncline. In the Pontian geosyncline, at the same time, the water depth increased and the flysch sedimentation began. After a short recession of the shoreline, a renewed advance took place during the late Campanian-Maestrichtian which marked the greatest expansion of the Cretaceous sea in Turkey.

As in earlier Mesozoic times, the Anatolian Tethys was divided into second order geosynclines and geanticlines. In a N–S profile the following units can be recognized:

1. Signs of a Pontian land are disappearing during the Cretaceous period. It is possible that the wildflysch detritus at Zonguldak-Cide still came from the north. During Late Cretaceous time, the land evidently was wholly submerged (*Muratov* and *Neprochnov,* 1967; *Brinkmann,* 1974: 69).

2. During Cretaceous time, the Outer Pontian geosyncline was fully developed, and in its entirety turned into a eugeosyncline characterized by flysch facies.

3. As in the Jurassic, Middle Anatolia represented a broad geanticlinal ridge which was furrowed by narrow and deep troughs. The ridge was subdivided into several domes – the Menderes, Kırşehir, and Bitlis Massifs. They were not flooded until Late Cretaceous time, and even then only partially. The most important trough was the Middle Anatolian ophiolite zone. It was growing to the west and east during the Cretaceous period. In the Ressen phase, the contents of the trough became mostly folded and uplifted. The continental and littoral-neritic sediments of the youngest Upper Cretaceous period are mainly composed of ophiolitic debris.

4. In the southern part of Anatolia, the miogeosyncline of the western Taurus remained intact all through the Cretaceous period. As in the Triassic, its inner part turned into an ophiolite trough, which probably was connected with the Southeast Anatolian trough via Cyprus. The ophiolite sequence in this zone continued into the Paleocene.

On the whole, during the Cretaceous period, the intensity of the epirogenic, tectogenic and magmatic processes was increasing. The sediment facies became more varied, and the paleogeographic map changed faster. Submarine volcanism was more widespread in the geosynclinal as well as in the shelf areas.

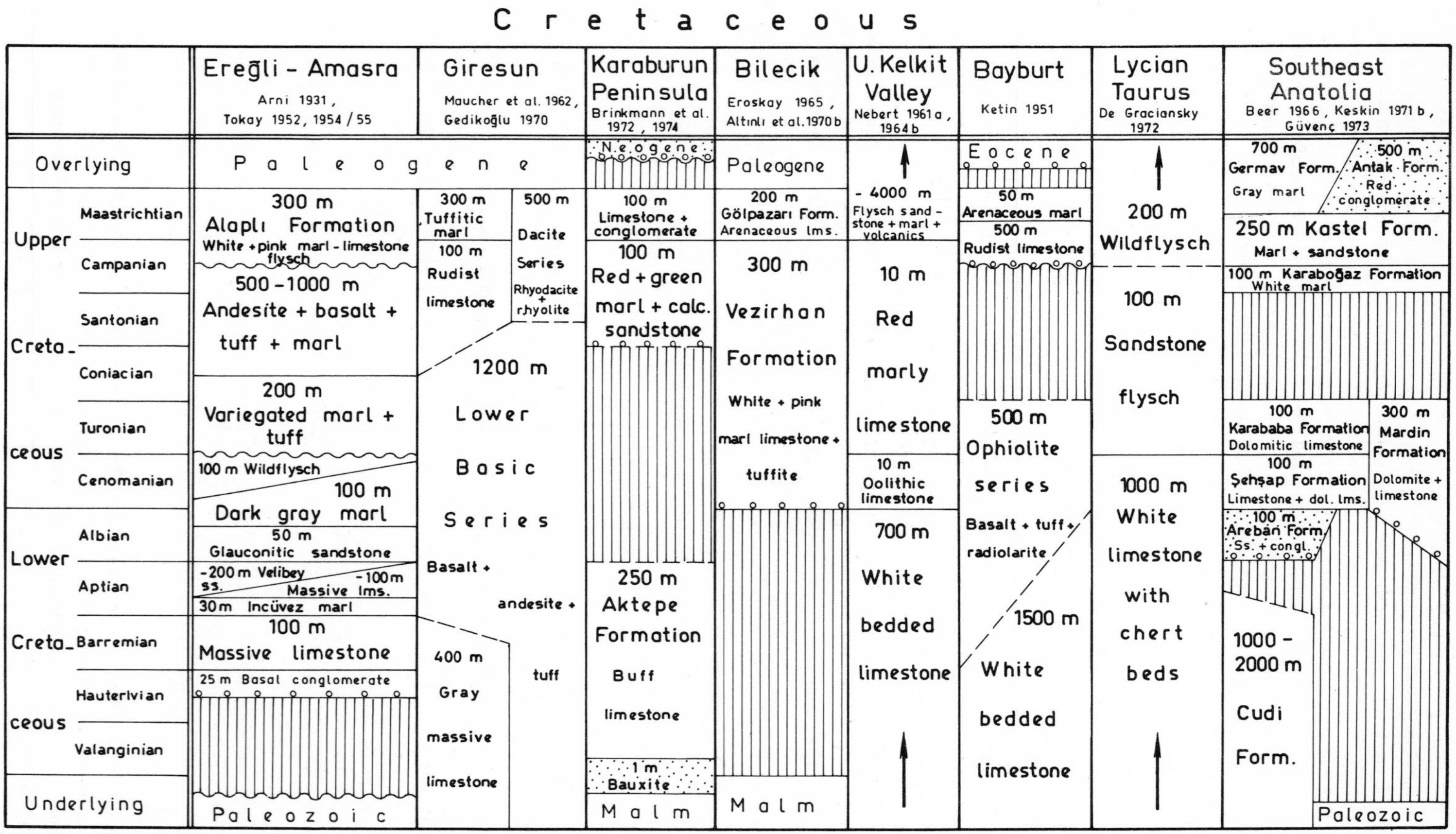

Cretaceous

Stages (left column): Overlying; Upper Cretaceous: Maastrichtian, Campanian, Santonian, Coniacian, Turonian, Cenomanian; Lower Cretaceous: Albian, Aptian, Barremian, Hauterivian, Valanginian; Underlying

Ereğli - Amasra (Arni 1931, Tokay 1952, 1954 / 55)
- Overlying: Paleogene
- 300 m Alaplı Formation — White + pink marl - limestone flysch
- 500 - 1000 m Andesite + basalt + tuff + marl
- 200 m Variegated marl + tuff
- 100 m Wildflysch
- 100 m Dark gray marl
- 50 m Glauconitic sandstone
- -200 m Velibey ss. / -100 m Massive lms.
- 30 m Incüvez marl
- 100 m Massive limestone
- 25 m Basal conglomerate
- Underlying: Paleozoic

Giresun (Maucher et al. 1962, Gedikoğlu 1970)
- Overlying: Paleogene
- 300 m Tuffitic marl
- 100 m Rudist limestone
- 500 m Dacite Series — Rhyodacite + rhyolite
- 1200 m Lower Basic Series
- Basalt + andesite + tuff
- 400 m Gray massive limestone

Karaburun Peninsula (Brinkmann et al. 1972, 1974)
- Overlying: Neogene
- 100 m Limestone + conglomerate
- 100 m Red + green marl + calc. sandstone
- 250 m Aktepe Formation — Buff limestone
- 1 m Bauxite
- Underlying: Malm

Bilecik (Eroskay 1965, Altınlı et al. 1970b)
- Overlying: Paleogene
- 200 m Gölpazarı Form. — Arenaceous lms.
- 300 m Vezirhan Formation — White + pink marl limestone + tuffite
- Underlying: Malm

U. Kelkit Valley (Nebert 1961a, 1964b)
- - 4000 m Flysch sandstone + marl + volcanics
- 10 m Red marly limestone
- 10 m Oolithic limestone
- 700 m White bedded limestone

Bayburt (Ketin 1951)
- Overlying: Eocene
- 50 m Arenaceous marl
- 500 m Rudist limestone
- 500 m Ophiolite series — Basalt + tuff + radiolarite
- 1500 m White bedded limestone

Lycian Taurus (De Graciansky 1972)
- 200 m Wildflysch
- 100 m Sandstone flysch
- 1000 m White limestone with chert beds

Southeast Anatolia (Beer 1966, Keskin 1971b, Güvenç 1973)
- 700 m Germav Form. — Gray marl / 500 m Antak Form. — Red conglomerate
- 250 m Kastel Form. — Marl + sandstone
- 100 m Karaboğaz Formation — White marl
- 100 m Karababa Formation — Dolomitic limestone / 300 m Mardin Formation — Dolomite + limestone
- 100 m Şehşap Formation — Limestone + dol. lms.
- 100 m Areban Form. — Ss. + congl.
- 1000 - 2000 m Cudi Form.
- Underlying: Paleozoic

At the same time, the present-day topography of Turkey began to take shape. In Middle Anatolia, a west-trending welt emerged from the fusion of the big crystalline massifs with the tectogene of the Middle Anatolian ophiolite zone – the first outline of the Anatolian mainland.

This area of uplift divided the Tethys into two straits, a fact of biogeographic consequences. During the Early Cretaceous epoch, all of Turkey belonged to the Mediterranean realm. During Late Cretaceous time, the Pontian geosyncline was populated by Central and East European animal communities, such as Inoceramus and Belemnitella, whereas the rudists remained characteristic for the rest of Anatolia (*Frech,* 1910: 21; *Nöth,* 1931: 357; *Stchépinsky,* 1942: 61).

Chapter 15 Tertiary

Knowledge of the system spread from Thracia, where *Boué* (1840), *Viquesnel* (1868), *Von Hochstetter* (1870), *Hoernes* (1876), *Calvert* and *Neumayr* (1880) were active. Next, the Anatolian Tertiary system was investigated, primarily by *Spratt* (1858), *De Tchihatcheff* (1866–69), *Schaffer* (1901–03), and *Philippson* (1910–15). So far, only a few groups of its rich fossil assemblages have been looked at: foraminifera by *M. Taşman* (1949), *Butterlin* and *Monod* (1969), *Dizer* (1972), and *Sirel* (1972); invertebrates by *D'Archiac* et al. (1866–69), *Blanckenhorn* (1890), *Schaffer* (1901–03), *Broili* (1911), *Daus* (1915), *Oppenheim* (1919), *Kühn* (1926), *Stchépinsky* (1939), *Pınar* (1954), *L. Erentöz* (1958), *Roman* (1960), *Sönmez-Gökçen* (1973), and *Yazlak* (1973); fishes by *Rückert-Ulkümen* (1965); mammals by *Ozansoy* (1965) and *Sickenberg* (1975a); pollen by *Akyol* (1964), *Nakoman* (1967) and *Benda* (1971); leaf flora by *Engelhardt* (1903). Summaries on the Tertiary system in Turkey have been compiled by *Chaput* (1936: 249), *Kleinsorge* (1961), *Alpan* and *Lüttig* (1971). *Lüttig* and *Steffens* (1976) are preparing a series of paleogeographic maps of the Oligocene and the Neogene in Turkey.

The Early Tertiary in Thracia (Fig. 21, 22)

Since Paleozoic time, the main part of Thracia has almost constantly been an area of denudation. During that period, old crystalline rocks have been exposed over vast areas. A subsidence started during the Middle Eocene epoch. The Thracian basement platform tilted at a fracture zone which roughly followed today's earthquake belt (p. 97). In the late Eocene, the transgression proceeded, reached the Istranca Massif, and finally the Rhodope Massif. It thus created a wide strait between the Mediterranean and the Black Sea.

During the Oligocene epoch, a delta advanced from Northwest Anatolia towards Thracia, and the marine basin became brackish and shrank. In the shallow sea, beds of manganese ore and on the land, lignite with mammal remains were deposited (*Ozansoy,* 1962; *Akyol,* 1964; *Nakoman,* 1966a). At the end of the Middle Oligocene epoch, the basin dried out. The connection between the Mediterranean and the Black Sea was interrupted again (*Kopp* et al., 1969; *Sönmez-Gökçen,* 1973).

The lithology of the lower Tertiary rocks in Thracia is uniform; fine-grained sandstone and sandy clay are prevalent. Some are flysch-like and have been deposited at considerable depth (*Gökçen* and *Ataman,* 1973). Others are deltaic, as evidenced by the intercalations of coal seams and gravel. Calcareous sediments, oyster and nummulite limestone with intercalated calcareous algae-coral reefs were deposited only at the coastal rim along the Istranca Massif (*Kemper,* 1966; *Keskin,* 1971a).

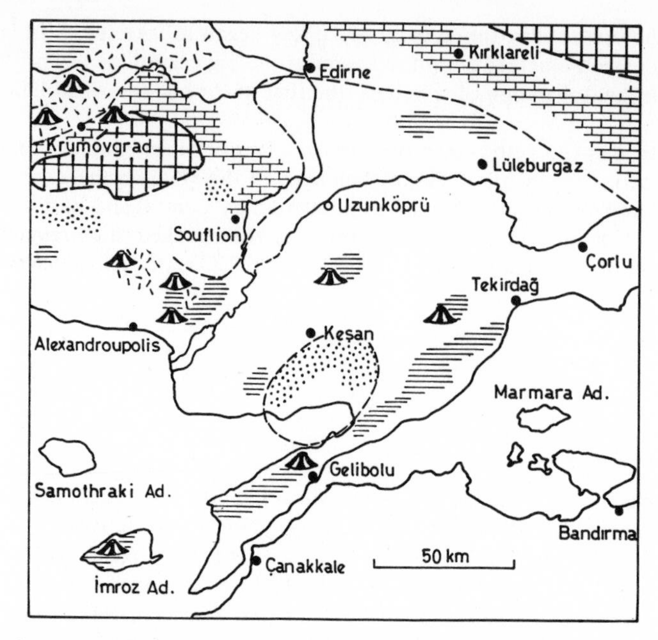

Fig. 21 Lithology and paleogeography of the Eocene/Oligocene transitional beds in Thrace (after *Ivanov* and *Kopp,* 1969a)

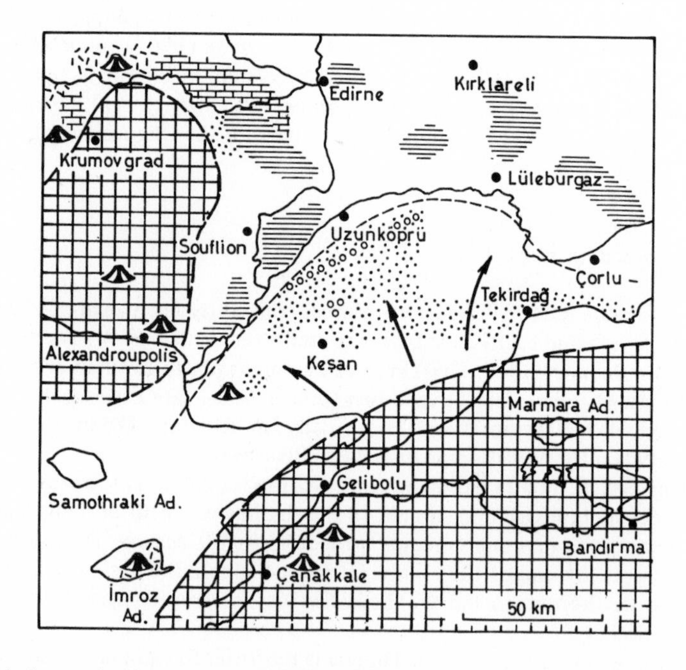

Fig. 22 Lithology and paleogeography of Middle Oligocene in Thrace (after *Ivanov* and *Kopp,* 1969a)
For explanation see fig. 21

The western and northwestern edge of the Thracian Basin was covered with volcanoes. Their mainly andesitic eruptive material is limited to the western areas, except for Canakkale, where it extends beyond the strait to Northwest Anatolia (*Ivanov* and *Kopp*, 1969a: 143).

The Paleocene and Eocene in North Anatolia (Fig. 23, 24)

The Outer Pontian geosyncline was present until the end of the Eocene epoch. From Cretaceous to early Tertiary time, it became the scene of increasing tectonic activities. Welts ascended which are characterized by incomplete sections, whereas the deposits in the basins continue without interruption from the Maestrichtian to the Paleocene (*Dizer*, 1972). Such processes are characteristic for the middle part of the geosyncline between Zonguldak and Samsun (*Blumenthal*, 1948: 135). The increased activity of crustal movements may have been one of the reasons for the change in facies. The limestone-marl flysch, which is prevalent during the Late Cretaceous epoch, is being replaced by a sandstone-siltstone flysch starting with the Paleocene. The submarine volcanism was decreasing in the Maestrichtian stage; during the Tertiary and particularly during the Eocene, it became active again. Andesite is the main material, dacite and basalt are less common. As in Cretaceous time, the eruptions increased from west to east. Sedimentary rocks are prevalent west of the Yeşil Irmak river, whereas east of it volcanites and tuffs are predominant in the Paleocene and Eocene formations (*Altınlı*, 1951a: 172; *Tokay*, 1954/55: 55; *Akartuna*, 1962b: 36; *Maucher* et al., 1962: 33, 66; *Gedikoğlu*, 1970; *Kilinç*, 1971; *Dizer*, 1953, 1956).

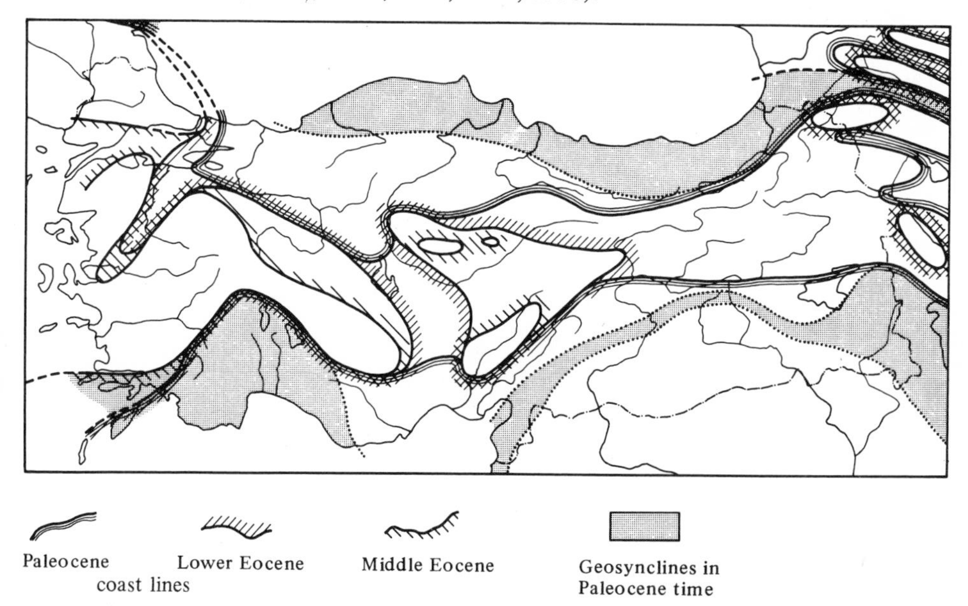

Fig. 23 Distribution of Paleocene and Eocene seas in Turkey (easternmost part after *Vinogradov* et al., 1961)

Towards the end of the Eocene epoch, the *Outer Pontian tectogene* emerged from the geosyncline. A more precise date for the age of the folding cannot given at present. The profiles end either with the Middle Eocene or the Late Eocene. It is probable that the Pyrenean phase was mainly active, but preliminary folding has also taken place.

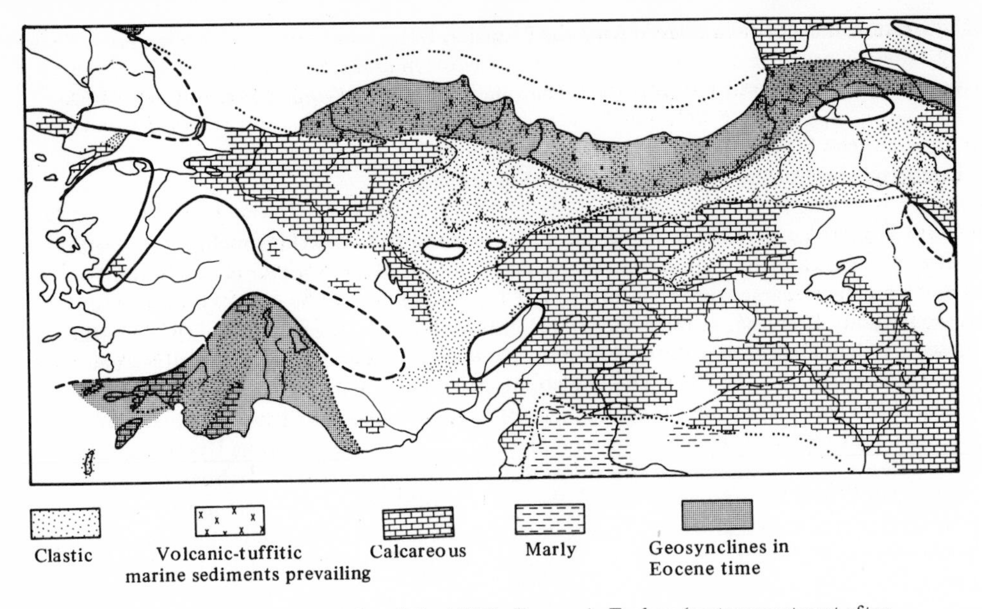

Fig. 24 Lithology and paleogeography of the Middle Eocene in Turkey (easternmost part after *Vinogradov* et al., 1961)

Efláni west of Kastamonu
(*Altınlı*, 1951a; *Baykal*, 1954)

Red beds (? Oligocene)
~~~~~~~~~~~~~~~
Middle Eocene
Lower Eocene
Paleocene
Maestrichtian
Lower Upper Cretaceous

South of Sinop
(Tidewater Oil Co., 1961)

Miocene
Volcanites
~~~~~~~~~~~~~~~
Upper Eocene
Lower Eocene
Paleocene
Cretaceous
Jurassic

Trabzon
(*Özsayar*, 1971)

Lower Pliocene
~~~~~~~~~~~~~~~
Upper Cretaceous

Kastamonu
(*Dizer*, 1953)

Neogene
~~~~~~~~~~~~~~~
Middle Eocene
Lower Eocene
~~~~~~~~~~~~~~~
Paleocene
Upper Cretaceous

South of Ordu
(*Gedikoğlu*, 1970; *Kilinç*, 1971)

Quaternary
~~~~~~~~~~~~~~~
Middle Eocene
~~~~~~~~~~~~~~~
Lower Eocene
? Upper Paleocene
~~~~~~~~~~~~~~~
Maestrichtian

The tectogenesis opened channels for rising magma and hydrothermal solutions. A number of granitic plutons formed in the anticlines and solidified under a rather thin cover. The most important one is the more than 100 km-long Tatos Massif in Northeast Anatolia. It shows contact metamorphism in Middle Eocene rocks and a radiometric age of 40 m. y. (biotite, K/Ar, *Delaloye* et al., 1972). The pyrite-copper ores are considered p. 103.

The Paleocene and Eocene in Middle Anatolia (Fig. 23, 24)

As a result of the tectogenic movements at the end of the Cretaceous, the central part of Anatolia became a west-trending welt which in the north sloped to the Pontian and in the south to the Taurian and Southeast Anatolian geosynclines. At the beginning of the Tertiary, it was mostly situated above sea level.

Between Bilecik and Ankara, the Paleocene series is composed of multicolored clastic strata with coal and gypsum lenses as well as interbedded marine layers. Andesite eruptions occurred occasionally. Toward the north, on Kocaeli and near Bolu, marine sediments become more frequent (*Stchépinski,* 1941; *Abdüsselâmoğlu,* 1959: 51; *Kalafatçıoğlu* and *Uysallı,* 1964: 6; *Eroskay,* 1965: 151; *Altınlı,* 1968: 15; *Altınlı* et al., 1970b: 81, 1972: 12). In Eocene time, the marine facies encompassed the entire area. The Lower Eocene series consists mostly of sandstone formations, the Middle Eocene of calcareous sandstone, marl, and limestone formations.

During the early Tertiary period, slight tectogenic movements took place:

Kocaeli
(*Baykal,* 1942; *Altınlı,* 1968)

Pliocene
∿∿∿∿∿
Middle Eocene
∿∿∿∿∿
Lower Eocene
Paleocene
Maestrichtian
Lower Upper Cretaceous

Bilecik
(*Eroskay,* 1965; *Altınlı* et al., 1970b)

Neogene
∿∿∿∿∿
Middle Eocene
∿∿∿∿∿
Upper Paleocene
∿∿∿∿∿
Lower Paleocene
Maestrichtian
∿∿∿∿∿
Campanian

Bursa-Gemlik
(*Erk,* 1942; *Altınlı,* 1943)

Neogene
∿∿∿∿∿
Oligocene
∿∿∿∿∿
Middle Eocene
? Lower Eocene
? Paleocene
∿∿∿∿∿
Maestrichtian
Lower Upper Cretaceous

West of Ankara
(*Kalafatçıoğlu* and *Uysallı,* 1964)

Neogene
∿∿∿∿∿
Middle Eocene
∿∿∿∿∿
Red beds (? Paleocene)
∿∿∿∿∿
Lower Paleocene
Maestrichtian
Lower Upper Cretaceous

Many radiometric dates, taken from granitic plutons, approximately coincide with the early Tertiary tectonic phases (*Çoğulu* et al., 1965; *Çoğulu* and *Krummenacher,* 1967; *Bürküt,* 1966; *Vachette* et al., 1968; *Ataman,* 1972, 1973). The most frequent ages are 60–55 and 40–35 m. y. In some places, the structural relationships argue for post-Cretaceous intrusions (*Norman,* 1972: 242); in other places, reheating processes may be anticipated.

In the early Eocene, the bay of Ankara-Haymana, which was formed during late Cretaceous time (p. 55), was enlarged to a strait that, running across Aksaray, connected the North Anatolian with the South Anatolian Eocene sea (*Yüksel,* 1970; *Norman,* 1973). The sea advanced also into the western part of Anatolia. At Izmir, nummulite limestone pebbles have been found in Miocene conglomerates, and north of Izmir, remains of Eocene marine deposits are present (*Nebert,* 1960: 21; *Akartuna,* 1962a: 8; *Dizer,* 1962a). Only the Menderes Massif apparently remained dry land; whereas the Kırşehir Massif subsided, except for a few islands, and was covered by sandy-calcareous sediments (*Arni,* 1938: 29; *Okay,* 1957: 58).

In many places of eastern Middle Anatolia, the Paleocene and Eocene deposits become thicker and coarser-clastic. Red and grayish-green conglomerates, consisting of radiolarite-ophiolite debris, extend from the uppermost Cretaceous (p. 57) to Paleocene and younger stages. Evidenced by gypsum lenses and land plant remains, they are characterized as continental to lagoonal facies (*Pamir* and *Baykal,* 1943: 314; *Ketin,* 1945: 292; *Baykal,* 1947: 201; *Kurtman,* 1961a: 2; *Demirtaşlı* and *Pisoni,* 1965: 30; *Norman,* 1972: 264, 1973a).

Upwards, the grain becomes finer and turns into pebble-bearing sandstone with Lower Eocene nummulites, but also with brackish fossils and, in places, even coal seams (*Arni,* 1938: 29; *Blumenthal,* 1950: 71; *Ketin,* 1951: 121, 1955b: 31; *Ozansoy,* 1957: 30; *Nakoman,* 1966b). The Middle Eocene rocks often overlap the Lower Eocene rocks unconformably and are more widespread than the older Tertiary series. They always show fully developed marine facies, either gray flysch sandstone or light-brown limestone which is composed of nummulite shells, coquina and algal nodules (*Blumenthal,* 1950: 69; *Nebert,* 1961a: 23; *Kurtman,* 1961a: 3, 1973a: 10; *Altınlı,* 1966: 48). Upper Eocene rocks of a sandy-marly facies are known only in a few places (*Blumenthal,* 1955: 91, 1958: 261; *Ayan* and *Bulut,* 1964: 65; *Akkuş,* 1971: 27). After the sea had reached its maximum extent in the Middle Eocene, a rapid regression happened.

Early Tertiary sedimentation was strongly influenced by epirogenic movements. Between Yozgat and Erzurum, a basin formed which followed the Middle Anatolian ophiolite zone and which collected more than 3000 m of Paleocene-Eocene deposits alternating with andesitic and basaltic volcanic rocks. Similar basins formed south of Erzurum and at the Van Gölü, but they did not subside more rapidly until the Neogene epoch (*Kurtman* and *Akkuş,* 1971; *Kalkancı,* 1974: 23).

The angular unconformities within the lower Tertiary system:

Yozgat
(*Ketin,* 1955b)

Red beds (? Oligocene)
Middle Eocene
Lower Eocene
Paleocene

Sivas
(*Kurtman,* 1961, 1970)

Aquitanian-Burdigalian
Red beds (? Oligocene)
Middle Eocene
Lower Eocene
Paleocene
Maestrichtian

Darende (east of Malatya)
(*Akkuş,* 1971)

Burdigalian
Upper Eocene
Middle Eocene
Volcanites
Maestrichtian

correspond, as in West Anatolia, with the Laramian and Pyrenean phases; however, they were generally more pronounced, compared to the ones to the west. Both phases were accompanied by the eruption of granitic magma. Several plutons between Kırşehir, Sivas, and Elaziğ are dated, radiometrically or by the contact relations, to be either post-Cretaceous/pre-Eocene or post-Eocene/pre-Oligocene (*Kovenko,* 1939; *Gysin,* 1943; *Ketin,* 1959b: 84, 1959c: 166; *Ayan,* 1963: 322; *Leo* et al., 1974; *Kalkancı,* 1974: 110).

The Paleocene and Eocene in South Anatolia (Fig. 23, 24)

Western Taurus

As in the Pontian geosyncline, epirogenic movements increased in intensity in the geosyncline of the western Taurus. Clastic sediments started to spread. The limestone and marl in the Lycian as well as in the Pamphylian basin were replaced by flysch sandstone and occasionally by wildflysch. Only the Bey Dağlari welt was not covered by sand; here calcareous strata were deposited until Oligocene time (*Brunn* et al., 1972a: 531; *De Graciansky,* 1972: 239; *Magné* and *Poisson,* 1974).

At the turn of the Cretaceous, the tectogenesis commenced in the western Taurus. It started in the Pamphylian trough and ended there even before Middle Eocene time (*Monod* et al., 1974: 117). Then, the eastern part of the mountain range formed. The tectogenic processes may have been of Pyrenean age, however, its age remains uncertain due to the lack of Oligocene strata. In the western part, the zone from the Marmaris Peninsula to Denizli, Burdur, and Isparta has also been affected by the Pyrenean phase, as evidenced by the unconformably bedded Oligocene serpentinite conglomerates at Denizli (*Becker-Platen,* 1970: 19) and the detritus of Mesozoic rocks in the Miocene series of Fethiye (*De Graciansky* et al., 1972: 562). The part of the western Taurus that was folded during Paleogene time is here termed the *Inner Taurides.* Only the Lycian Basin and the Bey Dağlari were not affected by tectogenic movements during the early Tertiary period.

Denizli-Acıgöl
(*Dizer,* 1962b; *Becker-Platen,* 1970)

Oligocene
~~~~~~~~
Upper Eocene
Middle Eocene
Lower Eocene
~~~~~~~~
Crystalline/Mesozoic

Eastern part of western Taurus
(*Monod* et al., 1974; *Bering,* 1971)

Pliocene
~~~~~~~~
Helvetian
Burdigalian
~~~~~~~~
Middle Eocene
Lower Eocene
Paleocene
Cretaceous

The Eastern Taurus

The eastern border of the Taurus geosyncline extended from Seydişehir to Anamur. North of Alanya, the Eocene rocks consist of bathyal radiolarian marl, whereas at the upper Göksu river, they consist of nummulite limestones. They are conformably passing into the Cretaceous system by marine Paleocene sediments (*Blumenthal,* 1951: 55, 1956: 14). The major folding of the Bolkar Dağ took place during the Pyrenean phase. At the same time overthrusting occurred in the eastern Taurus (*Özgül,* 1971: 98).

Bolkar Dağ
(*Blumenthal,* 1955)

Red beds (including Oligocene)
~~~~~~~~
Eocene
Paleocene
~~~~~~~~
Ophiolite stratum (including Maestrichtian)
Ultramafite

The Paleocene and Eocene in Southeast Anatolia (Fig. 23, 24)

At the beginning of the Tertiary, the development of the Southeast Anatolian ophiolite trough was completed. At Ergani Maden and Van, the Paleocene rocks still show an ophiolite facies (*Ketin,* 1950: 155; *Kıraner,* 1959: 39). At Siirt, they consist of red conglomerates, the same as in the Maestrichtian stage (*Tolun,* 1951: 85; *Altınlı,* 1954b: 48). During early Eocene time, the areal distribution of the conglomerates was particularly wide in connection with a renewed tectogenic phase. Starting with the Tertiary period, a foredeep formed south of the ophiolite zone in which thick layers of sediments became deposited north of Siirt (*Özkaya,* 1974: 68).

Elaziğ
(*Ketin,* 1947a)

Rudist limestone (Maestrichtian)
Ophiolite stratum (including Aptian)

Van Gölü
(*Kıraner,* 1959; *Demirtaşlı* and *Pisoni,* 1965)

Burdigalian
Aquitanian
Lower Eocene
Ophiolite stratum (including Paleocene)
Ophiolite stratum (including Maestrichtian)

North of Adıyaman and Diyarbakır
(*Rigo De Righi* and *Cortesini,* 1964)

Lower Miocene
Middle Eocene
Red beds (? Maestrichtian-Lower Eocene)
Ophiolite stratum (including ? Jurassic, Lower Cretaceous, Upper Cretaceous)

North of Siirt
(*Tolun,* 1951, 1953; *Özkaya,* 1974)

Quaternary
Upper Pliocene
Lower Pliocene
Miocene
? Oligocene
Middle Eocene
Red beds (? Maestrichtian-Lower Eocene)
Ophiolite stratum (including Campanian-Maestrichtian)

Hatay
(*Dubertret,* 1955; *Roloff,* 1972)

Burdigalian
Middle Eocene
Lower Eocene
Paleocene
Maestrichtian
Ophiolite stratum (including Upper Triassic, ? Cretaceous)

Toward the south, the foredeep shallowed to an epicontinental sea; its sediments reflect the tectonic events in the trough. The Cretaceous/Tertiary boundary lies in the middle of a thick sequence of gray marl in which sandstone plates appear with the Paleocene. The marl is overlain by fine-grained equivalents of the above-mentioned red conglomerates; toward the south, it merges into limestone with Lower Eocene foraminifera. During the Middle Eocene epoch, all of South Anatolia was covered by a shallow sea in which thick beds of fossil-rich, light-colored limestone were deposited. They were evidently connected with similar strata in Middle Anatolia beyond the Bitlis Massif. Then, a rapid drying-up process took place; Upper Eocene strata are known only near the Syrian border (*Tolun,* 1962: 232; *Altınlı,* 1966: 58; *Rigo De Righi* and *Cortesini,* 1964: 1927; *Güvenç,* 1973: 44).

The Oligocene in Anatolia (Fig. 25)

The Oligocene epoch in Anatolia was mainly continental; only limited areas were episodically covered by the sea.

The Northern and Southern Coast

Two localities of marine Oligocene deposits at Trabzon and west of Trabzon at Ünye (*Maucher* et al., 1962: 33; *Özsayar,* 1971: 5) show that the southern coast of the ancient Black Sea was almost identical with the present shore line.

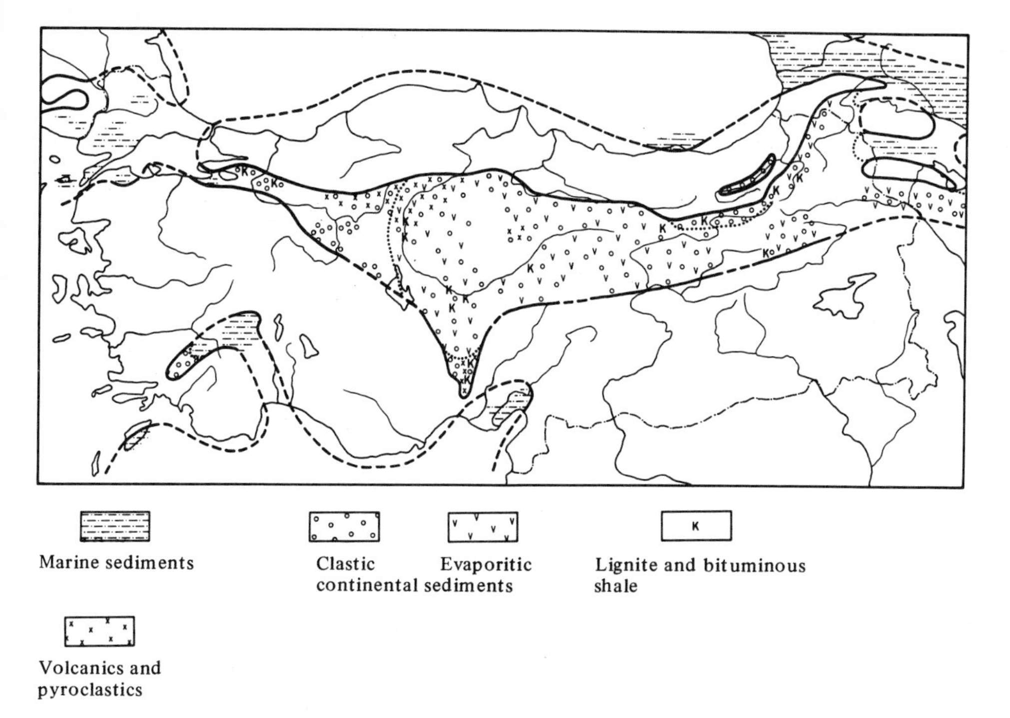

Fig. 25 Lithology and paleogeography of the Oligocene in Turkey (easternmost part after *Vinogradov* et al., 1961)

Two bays of the Oligocene Mediterranean Sea advanced into the present mainland, at Antalya and at Adana-Iskenderun. In the Muğla-Acıgöl-Dinar zone, 2000 m of multicolored and gray conglomerates and sandstones are exposed whose material originated in part from the Menderes Massif in the north and in part from the Taurus Mountains in the south. Because marine fossils are present in addition to limnic ones, this zone had an outlet to the Mediterranean (*Dizer,* 1962b; *Becker-Platen,* 1970: 19, 33; *De Graciansky* et al., 1972: 562). The connection must have been across Antalya, because marly limestones with Oligocene foraminifera have been found in the Bey Dağlari (*Magné* and *Poisson,* 1974). In the center of the Bay of Adana, bore hole samples have produced sandstone and multicolored marl with marine and brackish fossils (*Ternek,* 1957: 65). East of Adana in the Misis Dağlari, thick layers of a sandstone flysch of Oligocene-early Miocene age are exposed (*Schiettecatte,* 1971: 305).

In Southeast Anatolia, the Oligocene is missing over wide areas (*Güvenç,* 1973: 44). Only in the foredeep at Siirt, pillow lava with an age of 33–35 m. y. has been recently found. It is possible that the foraminiferal marl and the greenish flysch sandstone associated with the volcanites belong to the Oligocene series (*Özkaya,* 1974: 58).

Middle Anatolia

Multicolored conglomerates, sandstone, and sandy clay with thicknesses of up to 2000 m are common in Middle Anatolia. All rock types, from old crystalline rocks to Middle Eocene limestone, can be found as pebbles. Lenses and layers of gypsum are frequent, particularly in the environments of Sivas where they thicken up to 1000 m. Here too, was the center of halite precipitation, as judged by the occurrence of salt springs and diapir structures. Potassium-magnesia salt has not yet been found. On the other hand, the red beds contain also intercalations of lignite, bituminous shale, lacustrine limestone, and foraminiferal marl (*Lahn,* 1950; *Ketin,* 1955b: 32; *Kurtman,* 1961a, b, 1973a: 13; *Altınlı,* 1966: 50).

The major part of the sequence was deposited by intermittent streams and in ephemeral lakes. It accumulated in a basin that comprised large areas of Middle Anatolia and that collected the material eroded from the previously folded border mountains – the Outer Pontides in the north and the Inner Taurides and the Southeast Anatolian tectogene in the south. Uplift areas within the basin may also have contributed to the detritus. The thickness of the deposits was a function of the rate of subsidence; it was particularly rapid in the Yozgat-Erzurum zone (*Irrlitz,* 1972: 21, 88), already existing in the Eocene (p. 66). Here too, volcanism continued through Oligocene time. The basin had a seaward outlet to the east towards Armenia. The connection was temporarily wide open so that marine sedimentation could advance across the Kızıl Irmak to Cankırı in the west. But at most times, only a narrow inlet existed which caused the incoming waters to evaporate.

Due to the lack of fossils, the age of the red beds can only be estimated. At Darende west of Malatya, gypsum appears already in Upper Eocene strata (*Akkus,* 1971: 27). The lacustrine limestone, bituminous shale, and coal which are intercalated in the red beds have been classified as Oligocene (*Chaput,* 1936: 252 f.; *Blumenthal,* 1955: 93; *Akyol,* 1964: 45; *Beseme,* 1969; *Benda,* 1971: 24). According to *Kleinsorge* and *Vinken* (1965: 214) and *Kurtman* (1961), the sequence between the Kizil Irmak river and Sivas consists of

Marine Burdigalian-Helvetian
Upper red beds
Marine Aquitanian
Lower red beds
~~~~~~~~~~~~
Marine Middle Eocene.

The red beds, thus, are mainly Oligocene, but in places they may be as old as late Eocene, whereas their youngest portion is early Miocene.

The Neogene in Thracia and the Environments of the Straits (Fig. 26)

Lower and upper Tertiary rocks in the entire area are separated by a hiatus and an unconformity. Starting with the late Oligocene and lasting until the middle Miocene, a continental epoch existed in which a slight tectogenesis took place. At the same time, the direction of transport was reversed. The clastic material of the Upper Tertiary deposits originated from the north, from the denudation of the Rhodope-Istranca Massif (*Kopp* et al., 1969: 34).

Starting with the Helvetian stage, a new period of subsidence and flooding commenced. Variegated basal strata are overlain by yellowish, friable, thick-bedded calcareous sandstone, coquina, and finally oolitic limestone. They were deposited in a brackish sea that covered the Sea of Marmara and the Dardanelles regions and toward the north turned into a freshwater lake. During the Pannonian stage, this lake extended over major parts of southern Thracia. Also the surroundings of Istanbul and the Gulf of Izmit (*Akartuna,* 1968: 59) became drowned. West of Istanbul, coarse-grained sand with limnic-brackish molluscs and remains of continental and marine mammals, is unconformably overlying Eocene strata (*Chaput* and *Gillet,* 1939). The sand is succeeded by marl and coquina limestone with a brackish fauna, limited in species.

At the end of the Pannonian stage, the lake dried up. The area was covered with reddish sand carried by rivers from the Istranca Massif. In the late Pliocene, Thracia was leveled and transformed into the Thracian peneplain (*Cvijić,* 1908).
~~~~~~~~~~~~

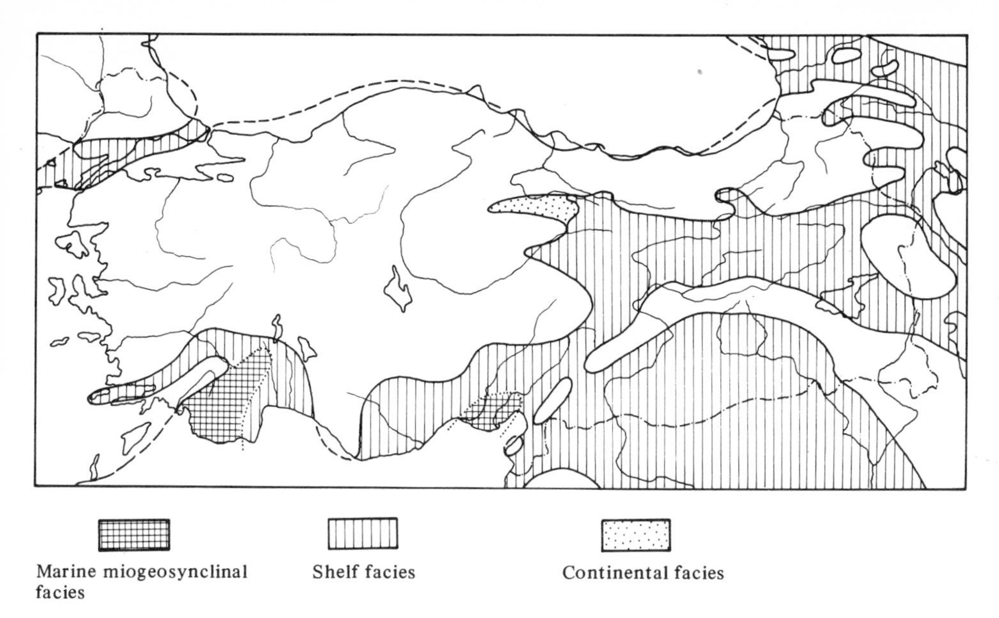

Fig. 26 Distribution, facies and paleogeography of the Middle Miocene in Turkey (after *L. Erentöz,* 1956, *Becker-Platen,* 1970, *Irrlitz,* 1972, *Vinogradov* et al., 1961)

The Neogene in North Anatolia (Fig. 26)

The epirogenic rise of the Pontian tectogene continued during the entire late Tertiary epoch. Sediments were deposited only on the northern and southern slopes of the mountain range, except for some intermontane basins.

The northern slope was affected several times by transgressions of the Black Sea; however, only in a narrow strip a few kilometers wide. Between Sinop and Giresun, sandy marl with a Burdigalian-Vindobonian marine fauna is present (Tidewater Oil Co., 1961). It is slightly unconformably overlain by sandy-humous clay with brackish-limnic Pannonian fossils (*Yalcınlar,* 1958: 13; *Maucher* et al., 1962: 33; *Irrlitz,* 1972: 59); these deposits are also present at Trabzon and Rize (*Özsayar,* 1971: 14). The sediments alternate here with andesites, basalts, and tuffs. Accordingly, the volcanism in the East Pontian Mountains lasted until approximately the end of the Pliocene period. So far, no marine Neogene sediments have been found on the southwest coast of the Black Sea. Only *Altınlı* (1968: 17) mentions reworked Miocene foraminifera on Kocaeli.

The Neogene in Middle Anatolia (Fig. 26)

The major part of Middle Anatolia was dry land during late Tertiary time, only its eastern part subsided temporarily below sea level during the Miocene. The general uplift and subsidence were accompanied by a warping of the crust which created basins and rises. These epirogenic movements, in conjunction with climatic changes, led to the formation of varied sediments.

Continental sediments were deposited over large areas. They consist of friable fanglomerates, conglomerates, and sandstones which, during arid periods, show a reddish-brown and during humid periods a grayish coloration. They alternate with lignite beds locally more than 100 m thick, and lacustrine marls and limestones. The marine sediments can be classified in gray or reddish, hard conglomerates, yellowish to brownish thick bedded calcareous sandstones, coquina, oolitic or algal-coral limestones, as well as in light-colored marls with deep-water gastropods. In addition, during regressive periods, gypsum marl and locally halite were deposited.

Lavas and tuffs covered large areas, mostly during the late Miocene and Pliocene epochs. They are alternating, in part, with the basin sediments, and in part, make up purely volcanic formations with thicknesses of more than 1000 m. Latite, dacite, and andesite are the major rock types, but rhyolite and basalt are also present. A considerable part of the tuffs is ignimbritic (*Westerveld,* 1957: 106).

Western Portion

At the beginning of the Neogene, Western Anatolia was an area of denudation; Oligocene and lower Miocene sediments are almost entirely missing. Then, a subsidence took place which progressed from west to east. In the Aegean coastal region, the sedimentation in the continental basins began with the Middle Miocene, and east of the Eskişehir-Afyon line, with the late Miocene or early Pliocene (*Becker-Platen,* 1970; *Bering,* 1971). The difference in elevation between west and east is also reflected in the fauna. According to mammal localities, Southwest Anatolia was a wooded area during late Miocene time, the environs of Ankara and Konya were a savanna (*Sickenberg,* 1975a: 50).

The late Tertiary Anatolian mainland reached far into the Aegean Sea toward the west. On the islands of Kos, Samos, Chios, and Lesbos, the same limnic sequences and mammals are present (*Besenecker* et al., 1968: 136; *Meulenkamp,* 1971; *Besenecker* and *Otte,* 1972). Only on the southern slope of the Kaz Dağ brackish Sarmatian molluscs are mentioned (*Van Der Kaaden,* 1959: 23).

The late Tertiary epoch in West Anatolia was a period of tectonic quiescence. The sequence in the continental basins is conformable almost everywhere (*Becker-Platen,* 1970: 194 f.). At the end of the Pliocene, major parts of the region were leveled into a lowland, age-equivalent to the Thracian peneplain. Its present elevation rises from 100–200 m on Kocaeli to 900–1200 m in the interior of Anatolia. Near Ankara, it truncates the top of the sequences of the Neogene. Thus, it can be dated as being of post-Pannonian age (*Erol,* 1973).

The northwestern part of Middle Anatolia was a center of Neogene magmatism. Several granitic intrusions with ages of 28 to 22 m. y. (*Di Paola* and *Innocenti,* 1969; *Leo,* 1972: 78) occurred whose impact may also have been registered in the Menderes and Kaz Dağ Massifs (p. 9, 11). In the area of Çanakkale, Balıkesir and Izmir, the volcanic eruptions were most active in the Middle Miocene and Lower Pliocene. Near Afyon and Konya, they reached their maximum at the turn of the Miocene to the Pliocene (*Fourquin* et al., 1970; *Steffens,* 1971: 48; *Bering,* 1971: 29; *Jung* and *Keller,* 1972; *Savasçın,* 1974; *Borsi* et al., 1973; *Benda* et al., 1974).

Eastern Portion

Eastern Middle Anatolia was subjected to an extensive transgression during Miocene time. In the Aquitanian stage, the sea advanced from Armenia and Iran beyond the Kızıl Irmak river in the west (p. 70), and to the north as far as the upper Kelkit valley and Erzurum (*Nebert,* 1961a: 30; *L. Erentöz* and *Öztemür,* 1964). The transgression probably reached its largest extent during the Burdigalian stage. At that time, an only 80 km-wide land barrier separated the East Anatolian from the Pontian Miocene sea. The sea receded in Helvetian time and left evaporation lagoons (*Demirtaşlı* and *Pisoni,* 1965: 31; *Afşhar,* 1965: 39;

Altınlı, 1966: 51; *Irrlitz,* 1972: 22). The regression was followed by a period of denudation and tectogenic movements. In late Miocene time, the areas of deposition were very limited. Not until the early Pliocene epoch did they begin to expand again and stayed till the early Pleistocene (*Staesche,* 1972; *Irrlitz,* 1972).

In the eastern part of Middle Anatolia, the Neogene crustal movements were stronger than in the western part. The Styrian folding is the youngest of the important phases. Upper Miocene and Pliocene strata are lying almost flat:

Merzifon, west of Amasya
(*Irrlitz,* 1972)

Lower Pleistocene
Upper Pliocene
Lower Pliocene
Tortonian
Red beds (? Oligocene)
Upper Eocene

Şebinkarahisar, west of Gümüşhane
(*Nebert,* 1961a; *Irrlitz,* 1972)

Lower Pleistocene
Upper Pliocene
Lower Pliocene
? Upper Miocene
Helvetian
Burdigalian
Aquitanian
Middle Eocene

Kayseri-Sivas
(*Lange,* 1971)

Lower Pleistocene
Pliocene
Upper Miocene
Burdigalian-Helvetian
Upper red beds
Aquitanian
Lower red beds
Middle Eocene

The volcanic activity, too, was stronger in the east than in the west. An eruptive center was situated at Nevşehir-Ürgüp during early Pliocene time (*Pasquaré,* 1968; *Lange,* 1971: 39). The surroundings of Erzurum-Kars were covered with lavas and tuffs during the Tortonian and again during the late Pliocene (*Irrlitz,* 1972: 91).

The Neogene in South Anatolia (Fig. 26)

At the southern coast of Anatolia, the Miocene transgression started from the Levantine Sea and advanced via the bays of Antalya and Adana-Iskenderun (p. 69). Its central parts had remained below sea level almost continuously since early Tertiary time.

Western Taurus

The Bay of Antalya widened as early as in the Aquitanian stage. A deeper basin formed in the area of the Lycian Taurus. From here the Muğla-Acıgöl-Dinar trough (p. 69) in the hinterland of the Taurides was again filled with water (*Becker-Platen,* 1970: 20). In the Lycian Basin sandstone flysch was deposited, the younger part of which turned into wildflysch. The flysch is thrusted by the rocks of the Lycian nappe (*De Graciansky,* 1972: 49).

Muğla-Acıgöl
(*Becker-Platen,* 1970)

Lower Pleistocene
Pliocene
Upper Miocene
~~~~~~~~~~~~~
Burdigalian-Helvetian
~~~~~~~~~~~~~
Aquitanian
Oligocene

Lycian Taurus
(*Becker-Platen,* 1970; *De Graciansky,* 1972)

Pliocene
~~~~~~~~~~~~~
Burdigalian
Aquitanian
Middle Eocene
Lower Eocene
Paleocene
Cretaceous

This tectogenesis which most likely was active in the Styrian phase, generated the *Outer Taurides.* Thus, mountain building as well as marine transgression in the western Taurus came to a close. During Pliocene time, several continental basins formed on the eroded mountain range; they were in part filled with salt lakes (*Becker-Platen,* 1970: 207; *Bering,* 1971: 23). A late Pliocene-early Pleistocene transgression, during which Kos and Rhodes were partially submerged, touched the Anatolian mainland only on the Marmaris Peninsula and at Antalya (*Chaput* and *Darcot,* 1953; *Orombelli* et al., 1967; *Becker-Platen,* 1970: 216; *Besenecker* and *Otte,* 1972).

Eastern Taurus

The bay of Adana-Iskenderun was more significant than the bay of Antalya in view of the extent and the duration of submergence. The Lower and Middle Miocene is marginally composed of marine conglomerates and coquina, and in the center of the basin of sandstone flysch with occasional boulders. Starting with the Tortonian stage, the bay began to fill up with sandstone, gypsum marl and fluviomarine gravel. However, a narrow coastal strip remained submerged until the Quaternary time (*C. E. Taşman,* 1950; *Ten Dam,* 1952; *Ternek,* 1953a, 1957; *L. Erentöz,* 1956: 18; *Schiettecatte,* 1971; *Rigassi,* 1971: 462). The 3000 to 5000 m-thick sequence was repeatedly folded:

Adana
(*Ternek,* 1957; *Schiettecatte,* 1971)

Quaternary
~~~~~~~~~~~~~
Pliocene
Upper Miocene
~~~~~~~~~~~~~
Helvetian
~~~~~~~~~~~~~
Burdigalian
Aquitanian
Oligocene

Volcanic rocks are missing except for a few tuff beds.

The northwestern borderland of the bay, the eastern Taurus, was flooded only during the Burdigalian and Helvetian stage. The sea remained shallow; today, the beds are still horizontal (*L. Erentöz,* 1956; *Akarsu,* 1960).

The Neogene in Southeast Anatolia (Fig. 26)

Following the Laramian tectogenesis of the Southeast Anatolian ophiolite zone, a foredeep had formed in its foreland. North of Siirt it subsided more rapidly. Here, thick layers of foraminiferal marl and flysch sandstone with conglomerate lenses were deposited which pass into multicolored gypsum marl and sand toward the foreland (*Özkaya,* 1974: 58).

In the course of the Pliocene, the sequence was subjected to renewed nappe movements (*Rigo De Righi* and *Cortesini,* 1964: 1935).

In the western part of the area, the foredeep remained shallower. At Gaziantep, Aquitanian or Burdigalian beds are overlapping Middle Eocene strata. Toward the top, the marine limestone is replaced by sandy gypsum marl and finally by reddish sand and gravel which came from the Bitlis Massif. Plateau basalts of late Pliocene-Quaternary age (p. 81) complete the profile and represent the only late Tertiary volcanic rocks (*Tolun,* 1962: 240; *Weber,* 1963: 681; *Altınlı,* 1966: 62). At the end of the Miocene epoch folding began and may have continued through the entire Pliocene.

Summary

During the Cenozoic period, the geologic development of Turkey progressed rapidly toward the present conditions. The progress was accentuated by several important tectogenic phases. Three stages can be distinguished:

1. The first stage, from Upper Campanian to Eocene time, set forth the tectonic and paleogeographic conditions of the Cretaceous period. The Outer Pontian eugeosyncline ran parallel to the northern coast of Anatolia. Toward the north, it grew shallow and turned into an epicontinental sea which reached across the present Black Sea to the Ukraine. Its east-west extension is confirmed by identical species of fish found in Thracia and at Erzurum (*Sönmez-Gökçen,* 1973a: 14). South of the eugeosyncline, in Middle Anatolia, the Ressen phase had created a geanticline, the major part of which was land in Paleocene time. During the Paleogene, it became subdivided into a series of basins and uplift areas. The deepest basins formed in the Tuz Gölü region and the eastern part of Middle Anatolia. The Menderes and Kırşehir Massifs represented the most important uplift areas which remained, at least partially, above sea level even during the Middle Eocene transgression. In South Anatolia, the geosyncline of the western Taurus and the Southeast Anatolian ophiolite trough existed until the early Tertiary.

   Then tectogenic movements commenced. The development of the Southeast Anatolian trough came to an end with the Laramian phase. The Pyrenean phase formed the Inner Taurides in South Anatolia. At that time, the Outer Pontides were folded along the Black Sea coast.

2. At the beginning of the second stage, during Oligocene time, the geographic map of Turkey resembled the present one in many aspects. The Mediterranean and the Black Sea were connected by a strait. To the north and south, Anatolia was surrounded by border mountains which embraced an inner area of subsidence comparable to the present Tuz Gölü Basin. At that time, however, the eastern part of Anatolia was topographically lower than the western part. The Oligocene red beds as well as the marine Miocene strata are almost entirely restricted to the east.

   During the Miocene epoch, probably during the Styrian phase, the unfolded part of the Taurus geosyncline was thrusted and the Outer Taurides formed. Crustal movements also occurred in Middle Anatolia. Here, as in Paleogene times, their intensity was a function of the thickness of the sediments.

3. During the third stage, lasting from late Miocene to the present, today's topography became established. West Anatolia subsided; one part was submerged beneath the Aegean Sea. Eastern Anatolia rose, and the Miocene sea receded. The uplift of Southeast Anatolia and northern Syria severed the connection between the Mediterranean Sea and the Indian Ocean. Thus the Tethys vanished in late Miocene.

Lower Tertiary (Paleogene)

	Thrace Kopp et al. 1969, Sönmez-Gökçen 1973	Zonguldak Bartın Tokay 1952, 1954/55	Giresun Maucher et al. 1962 Gedikoğlu 1970	Bilecik Eroskay 1965 Beseme 1969	Sivas Kurtman 1973	West. Taurus Brunn et al. 1972a	Southeast Anatolia Rigo de Righi + Cortesini 1964 Cordey 1971, Güvenç 1973
Overlying	Pliocene		Pliocene	Miocene	Miocene	Miocene	Miocene
Oligocene	250 m Pebble Form. 1200 m Lignite ss. 1000 m Marl Form. 50 m Fish shale		Limestone + tuffite	1200 m Red siltstone + sandstone with bituminous shale	1500 m Selimiye Formation Red sandstone with gypsum		? Germik Form. Marl + evaporites
Eocene Upper	600 m Keşan ss. 800 m Koru Dağ ss. 300 m Flysch ss. 100 m Reef lms.		300-500 m Youngest volcanics Basalt Trachyte + rhyolite			500 m Sandstone-marl flysch	300 m
Eocene Middle	100 m Reef + bedded limestone	– 2000 m Yellowish bedded sandstone	200 m Sandstone + marl	100 m Bedded Nummulite limestone	1000 m Köşedağ F. Flysch + Volcanics 1500 m Bozbel F. Flysch + tuffite		Midyat Formation Thick bedded limestone
Eocene Lower		150 m Glauconitic marl	200-1500 m Upper Basic Series		700 m Bahçecik Form. Conglomerate		800 m Antak Form. Red sandstone + conglomerate 250 m Gercüş F. Var. ss. 70 m Becirman limestone
Paleocene		150 m Light coloured limestone + marl	Andesite + basalt	250 m Kızılçay Form. Variegated clastics 50 m Selvipınar Formation White bedded lms.	500 m Gürlevik Formation Gray bedded marly limestone	20-50 m Cream + red marly Globigerina limestone	500 m Germav Form. Gray marl + sandstone
Underlying	Crystalline schist	Upper Cretaceous	Upper Cretaceous	Upper Cretaceous	Upper Cretaceous	Upper Cretaceous	Gray marl

Younger Tertiary (Neogene)

	Dardanelles Kopp et al. 1969	Istanbul Kopp et al. 1969	Southwest Anatolia Becker - Platen 1970	Central and Eastern Anatolia Bering 1971, Irrlitz 1972, Staesche 1972	Southeast Anatolia Rigo de Righi + Cortesini 1964 Güvenç 1973, Özkaya 1974	Adana Basin Ternek 1953 Schiettecatte 1971
Overlying					~250 m Flood basalt	~1000 m Gravel
Pliocene – Upper	200 m Thracian Form. Gravel + sand	50 m Belgrad gravel	50 m Milet Formation Bedded limestone	300 m Pisidian Form. / 400 m Pontian Form. / 250 m Elbistan Formation Freshw. sed. + lignite	100-400 m Lahti Formation Gravel + sand	500 m Coquina + marl
Pliocene – Lower (Pannonian)	100 m Ferrai Form. Greenish marl	100 m Küçükçekmece Form. Sand + marl + limestone	250 m Yatağan Formation Red + buff gravel + sand	Freshwater sediments + volcanics + lignite / Freshwater sediments	1000 m Congl. graywacke / 200 m Variegated arkose	300 m Marl + gypsum
Sarmatian	500 m Coquina + calc. sandstone		150 m Sekköy Formation Marl + limestone		Lice Form. Gray marl + sandstone / Şelmo Form. Variegated marl + gypsum	700 m Oyster sandstone + red siltstone
Miocene – Badenian – Tortonian			~20 m Main lignite seam			
Miocene – Badenian s. str.	50 m Variegated Basal Beds Red clay + conglomerate		200 m Turgut Formation Gray sand + gravel			
Miocene – Helvetian			100 m Calcareous sandstone	200 m Limestone + calcareous sandstone + marl Basal conglomerate	250 m Silvan Formation Thick bedded limestone + marl	1500 m Marly flysch
Miocene – Burdigalian						400 m Limestone + sandstone + marl
Miocene – Aquitanian			250 m Kurbalık Formation Sandstone + conglomerate	500 m Upper Red Beds / 50 m Marl + gypsum	100 m Gaziantep F.	Gray + red conglomeratic sandstone + marl
Underlying	Oligocene	Eocene	Oligocene	Lower Red Beds	Eocene	

A parallel development can be seen in the volcanism. Up to Eocene-Oligocene time, it was chiefly connected with the geosynclines. Starting with the Miocene, it began to spread areally across the country. The chemistry did not change significantly; intermediate calc-alkalic lavas prevailed throughout the Tertiary.

The Tertiary seas of Turkey as well as those of the Mediterranean region and the Near East were the habitat of a subtropical Tethyan fauna, which was especially rich during Middle Eocene time. Due to the occurrence of continental sediments, fossils of land organisms are more frequently encountered.

The late Miocene mammals of western Anatolia are related to those of southern Europe and in eastern Anatolia to those of Central Asia. At the beginning of the Pliocene, prairie and steppe dwellers of North American and Indian origin migrated into Turkey (*Sickenberg,* 1975a: 53). During the entire Tertiary period, the land flora resembled that of Europe. But some ties with the east also exist. East Anatolia, which became land after the Miocene sea had receded, received its plant cover from Central Asia (*Davis,* 1971: 20).

From the sediments and their floral and faunal content it becomes evident that temperatures have dropped since Eocene time. The proportion of subtropical land plants diminished during the Tertiary period, and in the late Pliocene was reduced to 5–10% (*Benda,* 1971: 23). Reef corals flourished in the Upper Eocene near Istanbul, in the Lower Miocene at Erzurum, and as late as in the Upper Pliocene at Marmaris, Rhodes, and Iskenderun. The amount of precipitation fluctuated, partly due to changes in topography. The Oligocene arid period has doubtless been enforced by the basinal shape of Anatolia, and the drought in the Early Pliocene by the drop of the Mediterranean water level (*Hsü* et al., 1973).

Chapter 16 Quaternary

The investigation of the Turkish Quaternary system started with four problem areas: the geomorphology (*W. Penck,* 1918), the formation of the Straits (*W. Penck,* 1919), the alpine glaciation (*Louis,* 1944), and the Paleolithic (*Pfannenstiel,* 1941). The Quaternary sediments have been investigated very little; more information is available on volcanic rocks. Also the fauna and flora have hardly been studied, except for the molluscs (*Archiac* et al., 1866/69), the early Quaternary mammals (*Ozansoy,* 1965; *Sickenberg,* 1975), and pollen (*Benda,* 1971). We owe a summary of the literature to *Erinç* (1970).

The Quaternary on the Mainland (Fig. 27, 28)

The Early Pleistocene

At the beginning of the Quaternary period, Thracia as well as Anatolia represented a peneplain with some isolated mountain ranges. The continental basins, in most of which subsidence continued from late Tertiary to Quaternary time, appeared as shallow depressions. Lithologically, the sediments resemble the Pliocene sediments; they can be distinguished only by their fossil content, particularly by the pollen flora and the mammal remains. At the beginning of the Quaternary, a steppe fauna of European character populated the region (*Sickenberg,* 1975a: 55, 1975b: 242). Man appeared at the same time, as evidenced by artefacts (*Ozansoy,* 1961: 53).

At approximately the middle of the early Pleistocene epoch, sedimentation ceased in all basins. Rivers began to erode the basin fillings and to dissect the plains (*Sickenberg,* 1975b: 244). This geomorphological transformation marked a tectogenesis which corresponds to the Wallachian phase. At the same time, the epirogenic subsidence of the West Anatolian coastal area and the rise of East Anatolia, which had been in progress since the later Tertiary, were accelerated. The consequences of these events are best revealed in the morphology of West Anatolia (*De Planhol,* 1956b; *Birot* et al., 1968).

The Wallachian tectogenesis also changed the river system in West Anatolia more than in East Anatolia. In the late Tertiary, the drainage from the highlands of the Menderes Massif and the Ulu Dağ was directed towards the north and the south. The early Pleistocene fault lines diverted the West Anatolian rivers into a westerly direction. The old drainage system remained intact only in Northwest Anatolia (*W. Penck,* 1918: 55). The river pattern of the central and eastern parts of Anatolia, however, has been inherited from the Tertiary. Frat (Euphrates), Murat, and Dicle (Tigris) followed the receding Helvetian sea. Yeşil Irmak, Kelkit, and Coruh must have existed as early as at the end of the Tertiary, because the ostracod fauna of the Northeast Anatolian Pliocene shows closer relations to the Pontian-Caucasian area than to the remainder of Anatolia (*Alpan* and *Lüttig,* 1971: 17).

With the uplift of Anatolia, the karstification increased (*Güldalı,* 1970). The Pisidian lakes at first had a surface, later a subsurface drainage. The water reappears near Antalya in the form of large springs (*Payne* and *Dincer,* 1965). Here, a 250 m-thick mass of calcareous tufa has been precipitated. Its age, which goes back as far as the middle Pleistocene epoch, gives a clue as to when the karst drainage might have started (*De Planhol,* 1956a).

The Middle and Late Pleistocene

During the ice age, the climate changed drastically several times. The glaciated areas in the Anatolian highlands increased during the glacial epochs. In the last glacial epoch, the Kaçkar Dağ in Northeast Anatolia, the most important glaciated region of Turkey, was bearing an almost 15 km-long glacier tongue ending at 1550 m above sea level. Today, it has dwindled to a length of 1–2 km (*Löffler,* 1970: 45). Also mountains that at present are free of ice, such as the Ulu Dağ and the Taurus, were at that time covered with glaciers (*Messerli,* 1967: 151; *Arpat* and *Özgül,* 1972). Almost all known glacier evidence dates back to the Würm, earlier glacial stages have left only doubtful remains (*Klaer,* 1965: 351; *Messerli,* 1967: 118). In all of Anatolia, the snowline during the last glacial epoch was 1000–1200 m lower than at present. Thus, the temperature in July must have been 6–8° C less (*Messerli,* 1967: 207). The glacial and the present snowlines run parallel to each other. Therefore, the climatic factors and the topographic conditions remained principally the same. Since the Younger Pleistocene, Anatolia's outline and elevation have not changed noticeably; the epirogenic uplift of the country has essentially come to a standstill.

Those parts of Anatolia that were situated below the snowline, were subjected to the effects of the periglacial climate during the ice ages. Fossil solifluction phenomena are repeatedly mentioned (*Löffler,* 1970: 82); loess has been proven by *Brunnacker* (1969). The climatic fluctuations were also felt in lower elevations. During the glacial epochs, the Anatolian lakes expanded due to reduced evaporation. Temporarily, the level of the Burdur Gölü rose by 90 m, the Tuz Gölü by 110 m, and the Van Gölü by 70 m. East of Konya existed a lake with an extent of 125 x 25 km which today has disappeared entirely. Its shore is characterized by shell-gravel terraces (*Louis,* 1938; *De Ridder,* 1965; *Bering,* 1971: 86; *Erol,* 1972). If the observations in Syria and other Mediterranean countries can be used

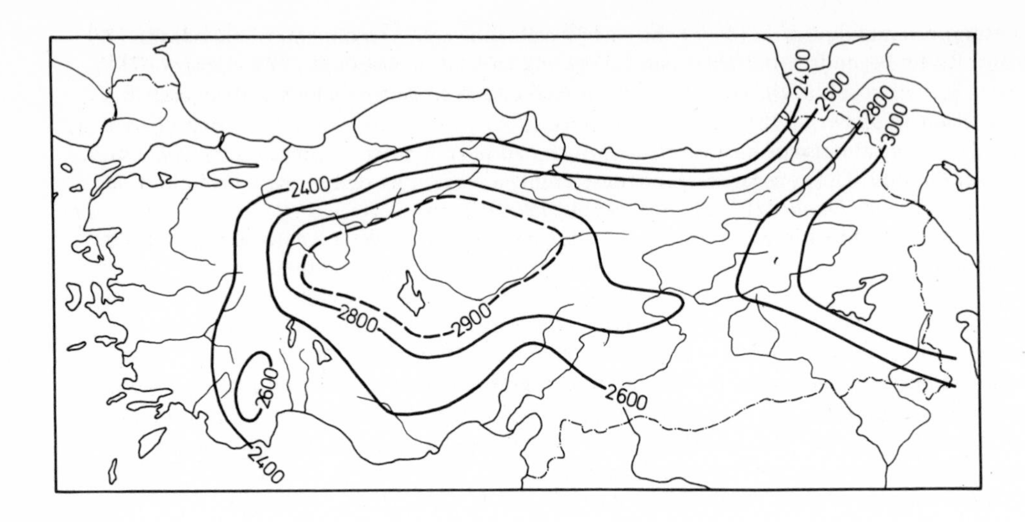

Fig. 27 Height of the snow line during the last ice age in Turkey (after *Messerli,* 1967)

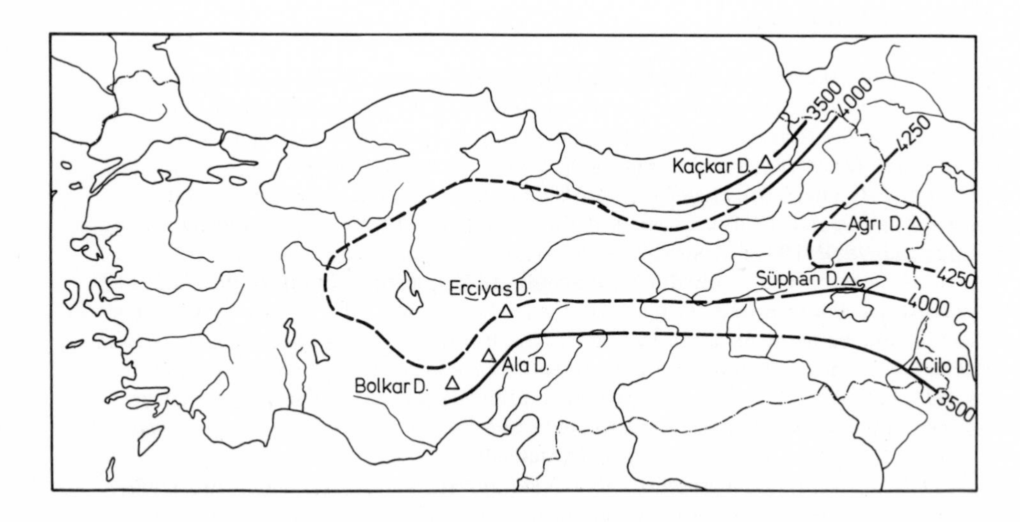

Fig. 28 Present height of snow line and distribution of glaciers (Δ) in Turkey (after *Messerli,* 1967)

for comparison (*Niklewski* and *Van Zeist,* 1970; *Wijmstra,* 1972), then Anatolia was covered by forests during interglacial times, and by a steppe during glacial times. The least favorable climate prevailed before the end of the last glacial epoch at approximately 17000 years B. C.

From the early Quaternary onward, the Anatolian mammal fauna is most closely related to that of Europe, and in particular to that of East Europe; the ties to Africa and Central Asia, which existed during the Neogene, were cut. Man appeared at an early stage (p. 78), but not until Younger Paleolithic time did tools become more frequent. Only teeth and footprints have been found of prehistoric man in Turkey (*Bittel,* 1950; *Müller-Karpe,* 1966; *Ozansoy,* 1969; *Barnaby,* 1975).

The Holocene

At the end of the last ice age, approximately 10000 years ago, the temperature rose rapidly to its present level (*Niklewski* and *Van Zeist,* 1970). Evaporation increased, and the groundwater level sank. Many springs ceased to flow, as shown by fossil calcareous tufa. Freshwater lakes turned into salt lakes or dried up entirely (*Irion,* 1973). During the Holocene epoch, the climate obviously did not change much (*Van Zeist,* 1972: 16), with the exception of the period 5500–2500 B. C., when the climate was somewhat warmer and more humid. At the turn of the Holocene, the inhabitants of Anatolia turned from hunting and gathering to a planned agricultural production by means of farming and ranching. A little later, at approximately 7000 B. C., they built the first townlike settlements at Konya and Burdur (*Hrouda,* 1971).

The Quaternary Volcanism

The early Pleistocene volcanism represented a continuation of the Pliocene volcanism. The same areas were affected: the central part of Anatolia from the Hasan Dağ north of Konya to the Erciyes Dağ near Kayseri as well as the eastern part from the Van Gölü to Erzurum and Kars. *Sanver* (1968) and *Durrani* et al. (1971) have obtained radiometric dates for the rocks of that area that are between 2.3 and 0.4 m. y. New fields of eruption came into existence at Kula west of Izmir (1.1 m. y.; *Borsi* et al., 1973: 475) and in Southeast Anatolia (1.04–1.45 m. y.; *Sanver,* 1968).

The volcanic cones of Turkey, and in particular the ones in East Anatolia, are mainly of Quaternary age (*Blumenthal,* 1958: 220; *Blumenthal* and *Van Der Kaaden,* 1964); however, detailed studies are missing. Judged by glacial morphology, the Erciyes Dağ was formed before and the Süphan Dağ after the last ice age. The Nemrut Dağ has dammed up the Van Gölü before the Riss ice age (*Klaer,* 1965: 353). Morphologically young calderas were formed in the western vicinity of the Erciyes Dağ and in the center of the Nemrut Dağ.

The youngest, well-preserved cinder cones at Kula east of Izmir (*Philippson,* 1913), in the Hasan Dağ (Fig. 1; *Jung* and *Keller,* 1972), as well as between Tuz Gölü and Erciyes Dağ are very probably of Holocene age; however, except Fig. 1 no records exist. In historic time, only the Acıgöl west of Nevşehir (*Pasquaré,* 1968: 140), the Erciyes Dağ (*Baykal* and *Tatar,* 1970), and the Nemrut Dağ west of the Van Gölü have been active. Solfataras occur near Isparta, in the Hasan Dağ, and in the Tendürek Dağ near the Van Gölü (*Westerveld,* 1957: 104). During the Quaternary, the eruptive material consisted mostly of basalt and locally of obsidian.

The Quaternary along the Coasts of Turkey (Fig. 29)

The Black Sea Coast

The North Anatolian coast has hardly changed since the folding of the Outer Pontides. Its eastern part represents an elevated shoreline in keeping with the epirogenic behavior of the Anatolian mainland. From 280 m on downward it is subdivided into several terraces, the lower ones are regarded as eustatic (*Löffler,* 1970: 102). They are all of Quaternary age, as they are cut into upper Pliocene basalt (*Özsayar,* 1971: 48). The coastal area situated between Sinop and Ünye, however, was temporarily submerged. Here, the Pliocene flooding continued until the early Pleistocene (*Yalcınlar,* 1958; *Irrlitz,* 1972: 61). West of Sinop there is again an elevated shoreline with terraces (*Nowack,* 1929: 5).

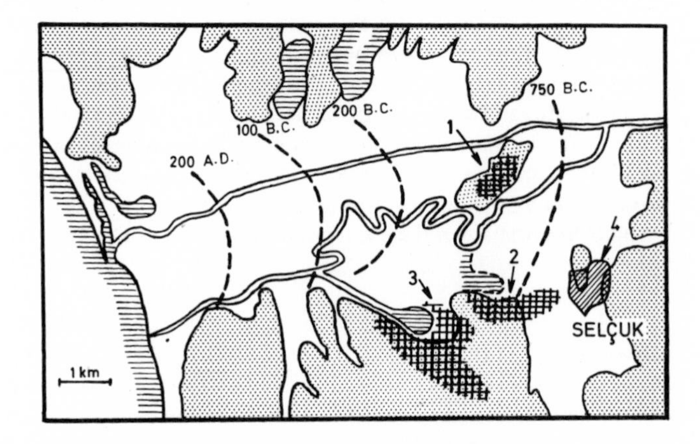

Fig. 29 Shift of coast line and settlements caused by the silting-up of the Küçük Menderes river, western Anatolia (after *Eisma,* 1962)

1 = Oldest settlement on an island in the bay, about 900–600 B. C.
2 = Ionian town Tracheia with harbour Koressos, about 600–300 B. C.
3 = Hellenistic-Roman town Ephesos with harbour, about 300 B. C. – 500 A. D.
4 = Byzantine-Seljukian town Selçuk as an interior settlement, since 500 A. D.

The Straits

During the Tertiary, the Mediterranean was repeatedly connected with the Black Sea by a strait via southern Thracia and the Marmara area. In the early Pleistocene, a renewed subsidence and transgression occurred, as can be discerned from the displacement of the Thracian peneplain. It forms the floor of the approximately 50 m-deep Marmara neritic sea and is downfaulted in the central Marmara trough to a depth of at least 1350 m. The banks of the Dardanelles and the Sea of Marmara are accompanied by a system of terraces that rise as much as 175 m. The level of its younger members coincides with that of the eustatic terraces in the Mediterranean and the Black Sea. The fauna in the terrace sediments and the fact that during the Pleistocene epoch Mediterranean molluscs repeatedly immigrated into the Black Sea, are further evidence for at least temporary connections between the two seas. Except for the terraces, the Dardanelles-Marmara region has been formed by fluviatile processes. These observations strongly suggest repeated fluctuations of the sea level. At low water, the Dardanelles were a river valley that drained the Sea of Marmara towards the Aegean Sea. During periods of transgression, they formed a strait (*Pfannenstiel,* 1944; *Akartuna,* 1968: 7; *Kopp* et al., 1969: 60; *Erol* and *Nuttal,* 1973; *Scholten,* 1974).

In contrast to the Dardanelles, the slopes of the Bosporus show a sequence of fluviatile terraces only. They were formed by a river which flowed from the Haliç (Golden Horn) into the Black Sea. It was not until the end of the last glacial epoch that the divide between the Haliç-Bosporus and the Marmara-Dardanelles river system was sufficiently low to be flooded by the eustatic rise of the Mediterranean. Accordingly, the present strait is young; during the earlier Quaternary, a strait presumably existed via Izmit to the lower Sakarya (*Pfannenstiel,* 1944).

The Aegean Sea Coast

The strongly dissected coastline is due to tectonic processes. Ever since the late Tertiary, West Anatolia showed a downward trend, and this subsidence was accelerated by faults. Early Pleistocene gravel plains are cut off at the present coast (Karaburun, unpub. data; Chios, *Besenecker* et al., 1968: 137). The graben of the Büyük Menderes river turned into a bay that extended as far as Aydin (*Ternek,* 1959). On the other hand, the presence of raised beach terraces is questionable (*Becker-Platen* and *Löhnert,* 1972). Also, the biogeographic conditions suggest that the land had once occupied larger areas in the Aegean, and that the Aegean islands among themselves, and with Anatolia and Greece were interconnected in various ways. Pleistocene mammal remains have been found on several islands, such as Kos, Kalymnos, Chios, and the Cyclades (*Besenecker* et al., 1972; *Kuss,* 1973). The distribution of the recent continental animals and plants also indicates the existence of land bridges, no matter whether they are of tectonic or of eustatic origin (*Rechinger,* 1950; *Wettstein,* 1953; *Strid,* 1971).

The tectonic subsidence still seems to continue today (p. 97). On the other hand, new land is being formed in the estuaries, particularly since the amount of debris carried by the rivers, has increased during the last millenia as a result of deforestation and agriculture. At the Büyük and Küçük Menderes, the average delta advance amounts to 6 m/yr.

The Levantine Sea Coast

Former land bridges were submerged also at the southern coast of Anatolia. According to mammal findings, Rhodes, Karpathos, Crete, and Cyprus were connected during the Pleistocene epoch with the mainland (*Kuss,* 1973: 61). The subsidence area of Adana-Iskenderun is still active. At the coast of Hatay, the partly marine Quaternary gravel beds show thicknesses of more than 1000 m (*Dubertret,* 1937: 111; *L. Erentöz,* 1956: 31; *Russell,* 1954: 382; *Erol,* 1969: 98; *Rigassi,* 1971: 462). Elevated terraces have been described from the middle part of the southern coast between Antalya, Anamur, and Silifke (*Ardos,* 1969).

Summary

The epirogenic crustal movements continued from the late Tertiary to the Quaternary period. East Anatolia was rising and West Anatolia subsiding. In places, the Miocene strata have been uplifted by 2000 m, and in places, they were submerged by 3000 m. The Quaternary tectogenesis followed a pattern different from the Neogene. Normal faults dissected Anatolia, and in particular its western part, into a blockland.

The tectogenic processes took place mainly during the early Pleistocene epoch; from then on, they have decelerated. The level of the late Pleistocene terraces and the height of the Würm snowline in Turkey fit well into the Mediterranean region. The volcanic activity, too, seems to have decreased in the course of the Quaternary.

Climatic fluctuations of the ice age forced the Pleistocene fauna and flora to repeated far-reaching migrations in which prehistoric man also took place. These migrations proceeded along paths which are no longer accessible.

Chapter 17 Alpidic Tectogenesis and Metamorphism

The Caledonian and Variscan mountain-building in Turkey took place only during a few phases and in a limited area. The Alpidic folding, in contrast, seized almost the entire country, although in varying manner and intensity. Its numerous phases continued throughout the Mesozoic and Cenozoic era. Obviously, it is still going on today.

The geosynclines of the Alpidic era were more varied than those of the Variscan one. During the Paleozoic era, almost only miogeosynclines were in existence. In contrast, the Mesozoic-Cenozoic geosynclines can be grouped into three types.

Alpidic Geosynclines

Miogeosynclines

Alpidic miogeosynclines are rare. Parts of the Taurus geosyncline show an almost complete sedimentary sequence of neritic facies. Also that part of the Outer Pontian geosyncline which is situated between Ereğli and Sinop remained free of volcanic activity from the Malm until the middle Cretaceous epoch.

Normal Eugeosynclines

This type was prevalent during Mesozoic and Cenozoic time. The Inner as well as the Outer Pontian geosyncline contain thick beds of flysch alternating with andesitic lavas and tuffs. The filling of the young troughs, which formed in East Anatolia during the Tertiary period, is of similar composition (p. 66, 70).

Ophiolite Eugeosynclines (Fig. 30, 31)

This type, which so far, has not been known in the Turkish Paleozoic, is distinguished from the above mentioned geosynclines by its fill and its basement. The fill, which here is named an *ophiolite sequence* (*Brinkmann,* 1968: 115), is composed of volcanic and sedimentary rocks that are partly unmetamorphosed and partly metamorphosed. The unmetamorphosed rocks consist of:

1. Spilite and diabase, more rarely keratophyre, in the form of veins, sills, and sheets; the latter sometimes show pillow structure. Originally, they frequently consisted of alkali basalt to trachyte (*Ayan,* 1963: 211; *Çoğulu,* 1967: 719; *Juteau,* 1970: 275; *Tusçu,* 1972: 28; *De Graciansky,* 1972: 333; *Gianelli* et al., 1972; *Piskin,* 1972: 80, 143; *Yazgan,* 1972: 70, 214).
2. Tuff and tuffite of the same type, modified by halmyrolysis (*Tusçu,* 1972: 21).

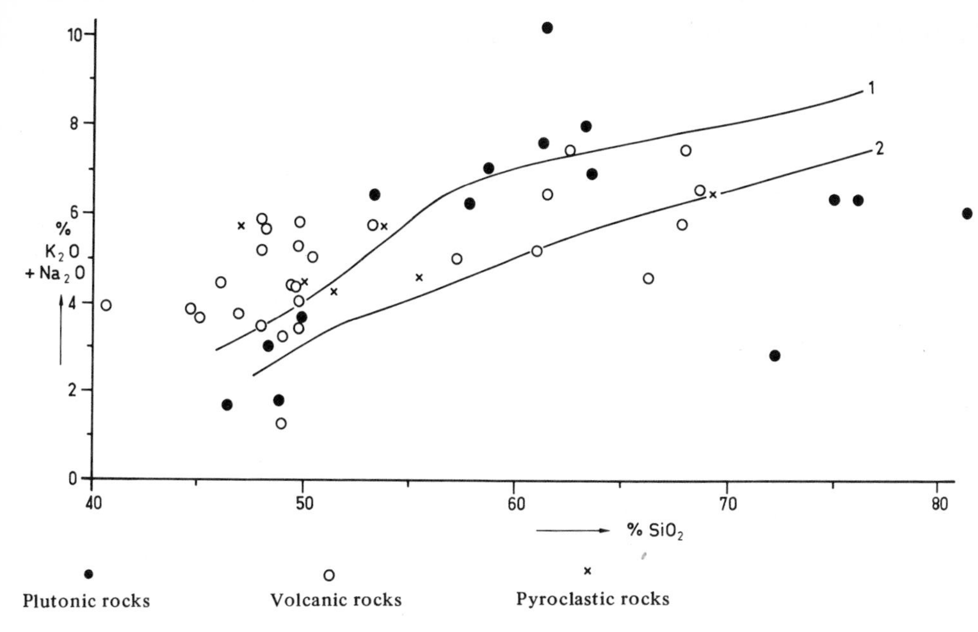

Fig. 30 Chemical composition of the Cretaceous/Paleocene magmatic rocks of the Southeast Anatolian ophiolite zone south and southeast of Malatya (after *Pişkin,* 1972; *Yazgan,* 1972)

1 = Limit between alkaline basalts (above) and Al-rich basalts
2 = Limit between the latter and tholeiitic basalts (below) (after *Kuno,* 1966)

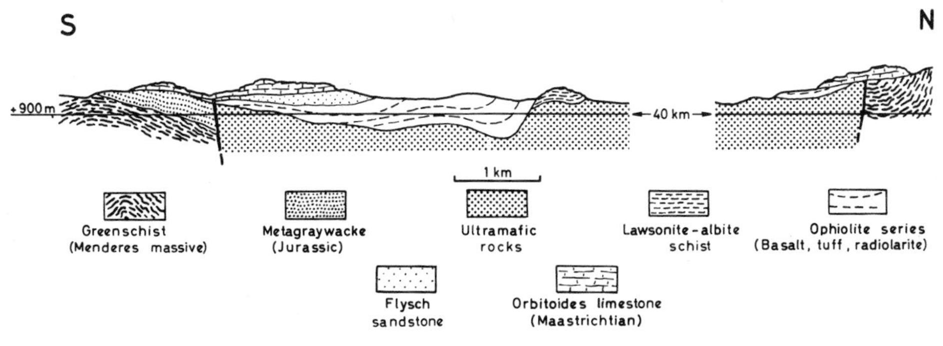

Fig. 31 Section across the Middle Anatolian ophiolite zone near Tavsanli west of Kütahya (after *Kaya,* mscr.)

3. Thin-bedded red and green radiolarite and siliceous shale, sometimes with beds of manganese ore.

4. Pink and light-colored, siliceous bedded limestone and fine-grained limestone with planktonic foraminifera (Globotruncana, Globorotalia) or ciliates (Calpionella). Like the radiolarite, they are abyssal sediments, as explained by *Garrison* and *Fischer* (1969).

5. Flysch sandstone, frequently with serpentinite and spilite grains (*Norman,* 1972, 1973a).

6. Debris of all grain sizes ranging from fine breccia composed of limestone, radiolarite, and spilite fragments, to huge exotic boulders of spilite and diabase, of Mesozoic and Paleozoic sedimentary rocks as well as of metamorphites, which alone or in series slumped into the tuffite or wildflysch (*Kaya,* 1972b: 495; *De Graciansky,* 1972: 239; *Norman,* 1972).

The metamorphosed part of the geosynclinal fill comprises:

1. Rocks of a basaltic appearence with ophitic structure in which plagioclase is replaced by lawsonite, and pyroxene by aegirine-jadeite or glaucophane.

2. Banded lawsonite-albite-glaucophane-actinolite schist which originated from basic tuff.

3. Thin-bedded piemontite-metaquartzite. It formed from radiolarite, as evidenced by the sporadic occurrence of fossils.

4. Marble as lenses and beds.

This rock assemblage obviously represents on ophiolite sequence, which was metamorphosed under the conditions of the lawsonite-glaucophane facies (*Çoğulu,* 1967: 740; *Kaya,* 1972a, b).

While the floor of the Alpidic miogeosynclines and normal eugeosynclines is formed by Paleozoic or by crystalline rocks, ultramafitites and mafitites compose the basement of the ophiolite sequences:

1. Harzburgite, more rarely dunite, forms massifs of more than 100 km, that border the adjacent rocks with steep tectonic boundaries. Their interior structure in the form of jointing, banding, chromite schlieren, and eruptive veins, is unconformably striking toward the rim of the massifs. Ultramafic veins are lacking in the adjacent rocks; thermal contact metamorphism is limited. The ultramafitite bodies of Anatolia belong to the „alpine“ type. They had already experienced a long history before they reached their present place as almost cooled, nearly solidified bodies (*Van Der Kaaden,* 1970; *Juteau,* 1970: 273; *Perrin,* 1970; *De Graciansky,* 1972: 421; *Engin,* 1972; *Gianelli* et al., 1972; *Aslaner,* 1973: 36; *Parrot,* 1973). Occasionally, amphibolite is associated with the ultramafitite (*Pişkin,* 1972: 51).

2. Gabbro and tholeiitic dolerite, more rarely intermediate magmatites, intersect the ultramafitite in the form of plutons or veins with distinct salbands (*Çoğulu,* 1967: 715; *De Graciansky,* 1972: 311).

According to observations in areas that have not been strongly deformed tectonically, the structural association of these three rock types can be shown in the following profile:

Bedded and reef limestone
Bedded limestone with chert nodules
Unmetamorphosed ophiolite sequence,
  at the top – flysch and wildflysch,
  at the bottom – lavas and tuffs prevalent

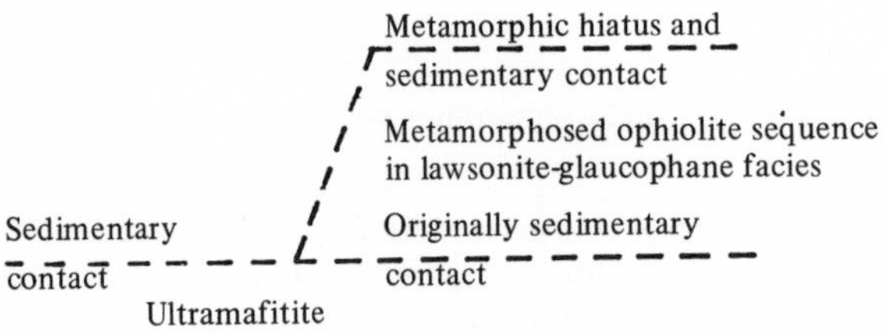

In not fully developed profiles, the unmetamorphosed sequence overlies the ultramafitite directly and may contain exotic blocks of the metamorphosed sequence (*Ketin,* 1950: 156; *Izdar,* 1963: 5; *Kaya,* 1972b: 495).

The dating of the ophiolite zones is based on the assumption that the emplacement of the ultramafitite bodies and the deposition of the ophiolite sequences are connected in time. The regular association of the ophiolite facies with the ultramafitites supports this. The biostratigraphic classification of the ophiolite beds is difficult due to the lack of characteristic fossils, the resedimentation processes and the partial metamorphism. So far, Upper Triassic, Jurassic, Cretaceous, especially uppermost Cretaceous, and Paleocene rocks have been found. The neritic sediments which cover the ophiolite sequences, are easier to classify; in some areas they belong to the Malm, in other areas to the Maestrichtian or the Middle Eocene.

Few radiometric dates are available. West of Ankara, K-Ar ages of 166–92 m. y. were obtained for gabbro and of 82–60 m. y. for lawsonite-glaucophane schist (*Çoğulu* and *Krummenacher,* 1967). Finally, the minimum age of the ophiolitic rocks is ascertained by the first appearence of their detrital relics. The oldest debris of lawsonite-glaucophane schist appears in the Liassic (*Erol,* 1954: 45) and Campanian stage (*Chaput,* 1936: 55; *Yüksel,* 1970: 39), and of serpentinite and spilite in the Malm (*Abdüsselâmoğlu,* 1959: 31; *Boccaletti* et al., 1966b). Starting with the Maestrichtian stage, ophiolitic debris becomes abundant. The ophiolite troughs in Anatolia, thus, seem to be limited to the Mesozoic and Paleocene. Many may have existed as early as in the Triassic. Each ophiolite zone had an individual lifespan; the ophiolite facies returned repeatedly (*Ketin,* 1951: 120; *Brinkmann,* 1972: 821).

Alpidic Tectogenesis

The Alpidic tectogenetic phases in Turkey fall into four groups according to their temporal and spatial distribution:

1. Early Alpidic group – Triassic to Jurassic
2. Middle Alpidic group – Cretaceous to Paleocene
3. Young Alpidic group – Eocene to Pliocene
4. Late Alpidic group – Late Tertiary to Quaternary.

Early Alpidic Tectogeneses (Fig. 32)

The early Alpidic folding was essentially limited to the Inner Pontian geosyncline. As the Liassic strata are often overlapping unconformably, early Kimmerian crustal movements are likely. The major phases were of middle and late Kimmerian age; they were responsible for the formation of the Inner Pontides. Little information is available concerning their structure. Probably west-trending open folds with a slight vergence to the north were formed (*Ketin,* 1951: 123). The rise of several granodiorite plutons accompanies the tectogenesis (p. 46 f.).

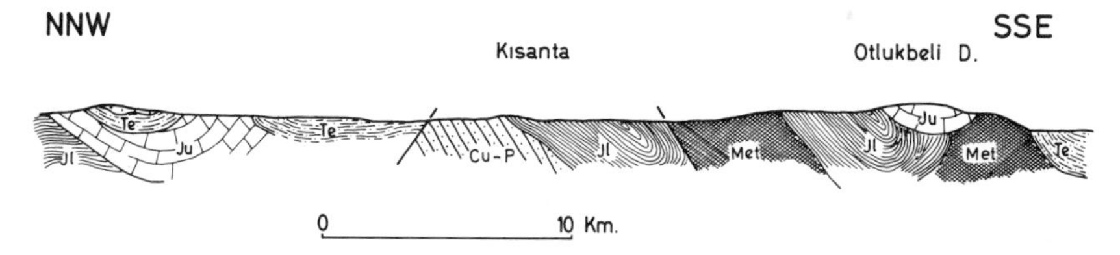

Fig. 32 Section across the Inner Pontides west of Bayburt (after *Ketin,* 1951).
Met = Crystalline schist, C–P = Carboniferous/Permian, Jl = Liassic, Ju = Malm, Te = Eocene

Outside the Pontian geosyncline, gaps in the geologic record and slight unconformities in the Triassic and Jurassic sequences of Kocaeli, Middle Anatolia, Karaburun, and Chios are attributable to the Kimmerian phases.

Middle Alpidic Tectogeneses (Fig. 33, 34)

The first weak movements occurred in middle Cretaceous time. One of them took place between the Cenomanian and Turonian stages and belongs to the Austrian phase. It was of certain importance in the Outer Pontian geosyncline.

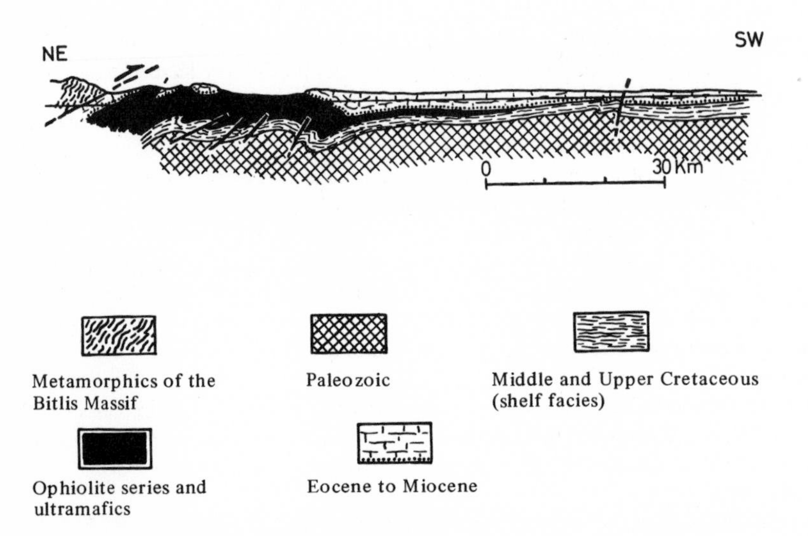

Fig. 33 Section across the Southeast Anatolian ophiolite zone near Kahta north of Adıyaman (after *Rigo De Righi* and *Cortesini,* 1964)

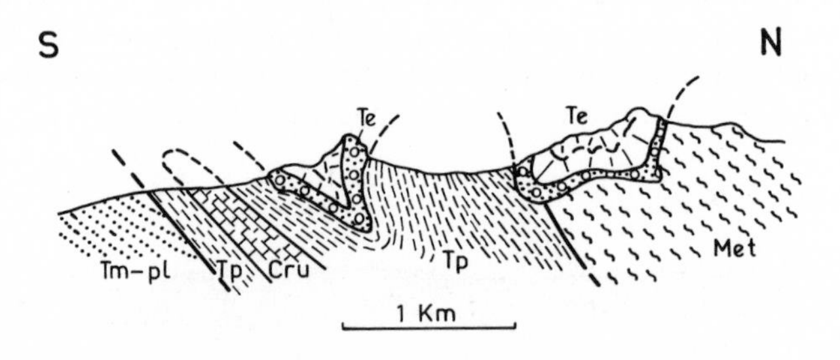

Fig. 34 Section across the front of the Bitlis nappe. Two phases of thrusting, pre-Eocene and post-Pliocene. Sason west of Bitlis (after *Tolun,* 1951)
Met = Metamorphics of the Bitlis Massif, Cru = Upper Cretaceous, Tp = Paleocene, Te = Eocene, Tm–pl = Miopliocene

Major folding took place during the end of the Cretaceous and the beginning of the Tertiary mainly affecting the ophiolite troughs. Wide areas of the Middle Anatolian trough were subjected to the Ressen phase, whereas its western part and the Southeast Anatolian trough were seized by the Laramian phase. At that time, the ophiolite sequences received their chaotic structure which earned them the terms „facies mixte ou brouillé" (*Blumenthal,* 1948) and „Ankara mélange" (*Bailey* and *McCallien,* 1954). The plastic serpentinite forms diapirs, fills breccia dikes and injects into thrust planes. The thin-bedded radiolarite is intensely folded. Areally extended structures and clear vergences are missing.

In the western Taurus, the nappes of Antalya are of Laramian age and consist mostly of ophiolitic rocks (*Lefèvre* and *Marcoux,* 1970). Their origin and geotectonic position – whether they are externides or internides – are controversial (*Brunn,* 1974; *Monod* et al., 1974). *Baykal* and *Kalafatçıoğlu* (1973: 40; *Kalafatçıoğlu,* 1973: 71) do not believe in nappes near Antalya and explain the structural relations by upthrusts.

In Southeast Anatolia, the ophiolite sequence and its ultramafic base have been seized by the middle Alpidic tectogenesis. According to *Rigo De Righi* and *Cortesini* (1964: 1928), this complex has been thrusted by the Malatya-Bitlis block and became transported towards the southern foreland partly as nappe and partly as olistostrome. Between Adıyaman and Siirt, the thrust is at least 15 km wide (*Arni,* 1940: 557; *Radelli,* 1971: 133). Very likely even more than that, for near Adıyaman, metamorphosed Permian and unmetamorphosed Cambrian rocks are within close proximity to each other (p. 96). Some of the Cretaceous deposits within the Bitlis crystalline Massif may well represent tectonic windows (*Baykal,* 1950: 149; *O. Yılmaz,* 1971: 191; *Hall* and *Mason,* 1972). The nappe movement took place during several phases – in the Campanian, in the Paleocene, and in the Neogene (p. 59, 67, 75). During this time-span the direction of compression changed. At the western edge of the Bitlis Massif, two groups of thrust and foliation planes cross each other (*Aykulu,* 1971: 47; *Aykulu* and *Evans,* 1974: 303). The amount of thrusting is decreasing toward Hatay. The Amanos Dağ represents an autochthonous, southeast-verging anticline (*Schwan,* 1972: 143). The ultramafitite massif of Iskenderun, however, has been thrusted. The Upper Cretaceous limestone underlying the ultramafitite is not its normal base (*Dubertret,* 1955: 102), but appears in a window or semi-window (unpub. data). These relations could explain, why the gravimetric effect of the ultramafic massifs is negligible (*Bourgoin,* 1945/48).

Toward the end of the Cretaceous, some large fracture zones became active in Middle Anatolia. A fault, striking parallel to the northeastern shore of the Tuz Gölü, separated the Kırşehir Massif from the Lycaonian Massif (p. 10) which may form the basement of the Tuz Gölü Basin (*Erol,* 1970: 40). The Ecemiş corridor (*Blumenthal,* 1941: 71) represents a pre-Eocene, left-lateral strike-slip fault of approximately 75 km offset (*Metz,* 1939b: 333) which points toward the Erciyes volcano in the north. During a late Laramian phase, Paleozoic strata on either side of the Bosporus have been thrusted northwards upon Upper Cretaceous rocks (*Chaput,* 1936: 152; *Akartuna,* 1963; Fig. 11).

Young Alpidic Tectogeneses (Fig. 35–44)

The Outer Pontides and the Taurides formed during the Pyrenean and Styrian phase. So the last of the major Mesozoic geosynclines became folded.

1. It seems that the Outer Pontides in their entire length were folded in the Pyrenean phase. The different parts do not show the same structure. The area from Ereğli to Bartın is characterized by open, slightly north-verging anticlines exhibiting the Variscan basement (*Tokay,* 1952: 1954/55). Between Bartın, Inebolu, Kastamonu and Amasya, overthrusting developed parallel to the bedding planes, especially at the base of the Malm limestone complex (*Fratschner,* 1954: 218; *Tokay,* 1962: 9; Tidewater Oil Co., 1961; *Alp,* 1972: 91). Farther east, at Giresun

and along the Coruh river, gently rolling folds are prevalent which are cut by numerous, mainly diagonally-trending faults (*Maucher* et al., 1962: 11, 88; *Kraeff,* 1963: 51; *Kronberg,* 1970). Plutons occupying the cores of anticlines are intercalated (p. 64). The Outer Pontides are mostly north-verging. But the southern marginal zone is folded to the south in certain regions, such as at Boyabat south of Sinop or near Şebinkarahisar south of Giresun (Tidewater Oil Co., 1961; *Nebert,* 1961a: 34).

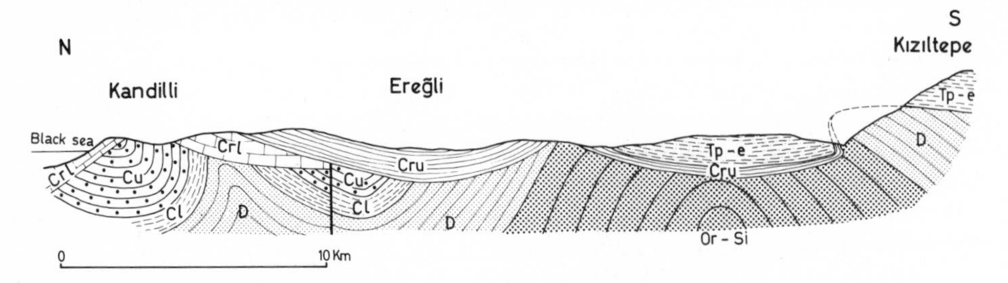

Fig. 35 Section across the Outer Pontides near Ereğli. Variscan and Alpidic mountain building (after *Tokay,* 1952)
Or-Si = Ordovician/Silurian, D = Devonian, Cl, Cu = Lower and Upper Carboniferous, Crl, Cru = Lower and Upper Cretaceous, Tp-e = Paleocene/Eocene

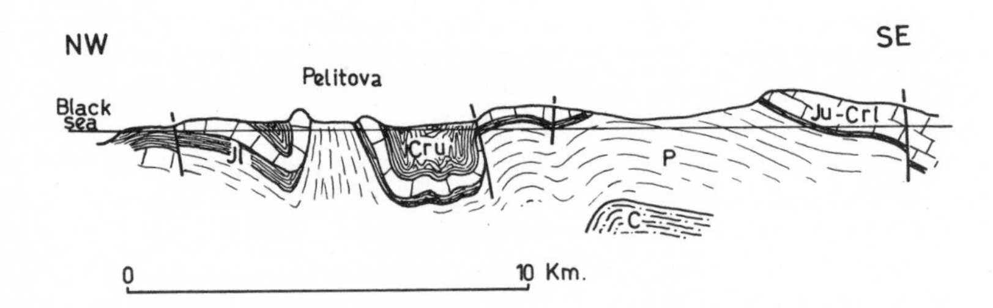

Fig. 36 Section across the Outer Pontides near Cide west of Inebolu. Kimmerian angular unconformity, main folding post-Cretaceous (after *Grancy,* 1939)
C = Carboniferous, P = Permian, Jl = Liassic + Dogger, Ju-Crl = Malm/Lower Cretaceous, Cru = Upper Cretaceous

2. The last tectonic movements in the Taurides are also young-Alpidic; they are divided into two stages. The Pyrenean phase folded the internal zone of the Taurus geosyncline into the Inner Taurides, and the Styrian phase folded the remainder of the geosyncline into the Outer Taurides. As first recognized by *Blumenthal* (1944b), the older phase created southwest-verging nappes southwest of the Sultan Dağ. *Gutnic* et al. (1968) regard the 150 km-long zone of the Hoyran Gölü-Beyşehir Gölü as a klippe. So far, the root problem remains open. Between Afyon and Isparta, the Inner Taurides turn to the southwest. At Denizli and Fethiye, pre-Oligocene movements have been shown (p. 67). The overthrust structure of Rhodes is also of Pyrenean age (*Mutti* et al., 1970: 166). The Styrian phase formed the Lycian nappe which thrusted the Bey Dağlari from the northwest. Its root is unknown. *De Graciansky* (1972: 567) indicated that it may have come from an area northwest of the Menderes Massif.

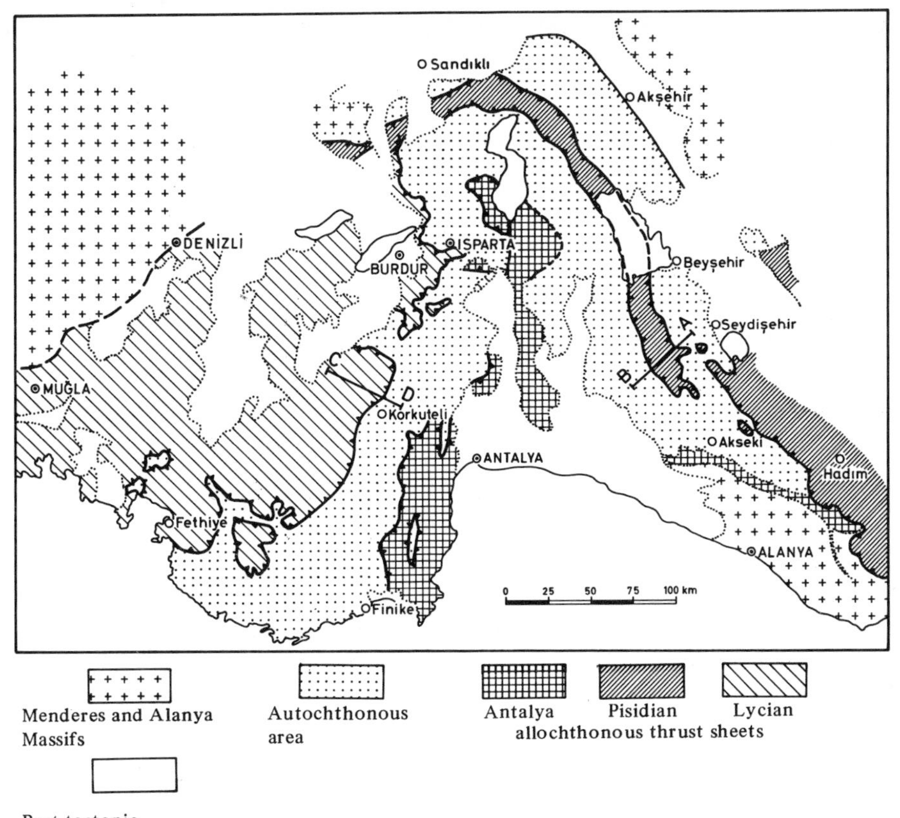

Fig. 37 Tectonic map of the western Taurus (after *Brunn,* 1974)

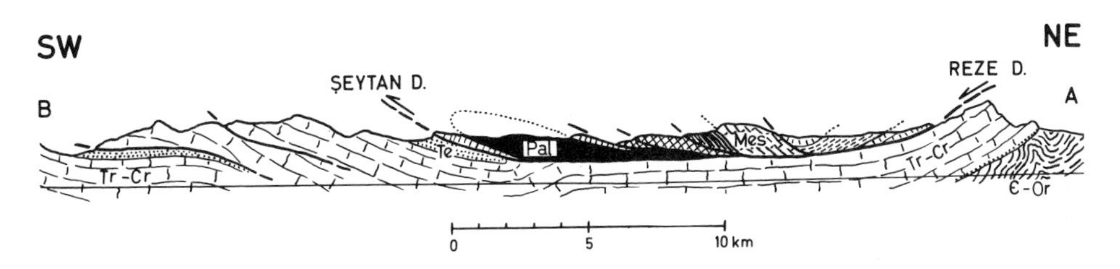

Fig. 38 Section across the Hoyran nappe near Seydişehir, Pisidian Taurus. Location of section see Fig. 37 (after *Brunn* et al., 1972a)
C-Or, Tr-Cr, Te = autochthonous Cambrian-Ordovician, Triassic-Cretaceous, Eocene.
Pal, Mes = allochthonous Paleozoic and Mesozoic

3. The Alpidic mountains in Southeast Anatolia have certain features with the Taurides in common, but their temporal and structural development is noticeably different. After the Laramian thrust in Southeast Anatolia, tectonic processes did not start again till the Pliocene. On the one hand, the movements in the overthrust zone revived extending at least another 15 km (*Rigo De Righi* and *Cortesini,* 1964: 1933; *Radelli,* 1971: 138; *Yılmaz,* 1971: 191).

On the other hand, in the southern foreland, a belt of regular, south-verging folds of the type of the Swiss Jura Mountains was created; it gradually vanished to the south. Folding evidently lasted through the entire Pliocene epoch and possibly the Quaternary (*Tolun*, 1953: 110; *Altınlı*, 1966: 5). According to *Yazgan* (1972: 57) the northern edge of the Bitlis Massif is a mirror image of the souther edge. East of Malatya, the ophiolite series and its Eocene cover show north-verging folds.

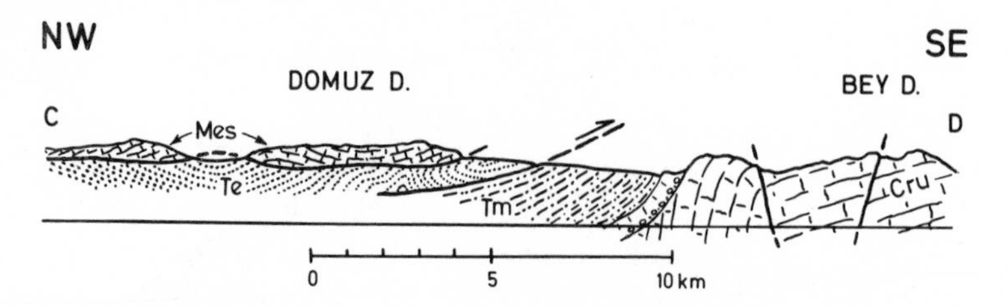

Fig. 39 Section across the Lycian nappes and the autochthonous Bey Dağlari. Location of section see Fig. 37 (after *Brunn* et al., 1972a)
Cru, Tm = autochthonous Upper Cretaceous and Aquitanian/Burdigalian,
Mes, Te = allochthonous Mesozoic and Eocene

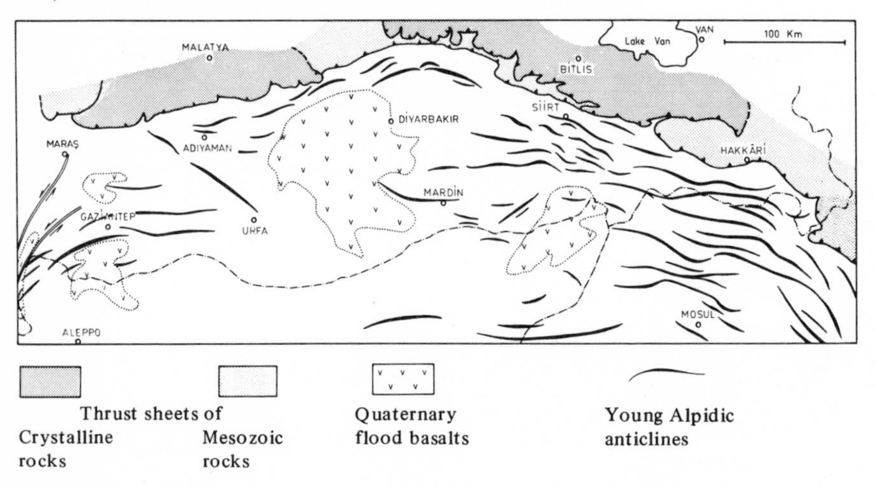

Thrust sheets of
Crystalline rocks | Mesozoic rocks | Quaternary flood basalts | Young Alpidic anticlines

Fig. 40 Foreland folds in Southeastern Anatolia (after *Tolun*, 1962; *Rigo De Righi* and *Cortesini*, 1964; *Altınlı*, 1966; *Wolfart*, 1967a)

4. The present structure of Middle Anatolia has principally been formed during the Pyrenean and Styrian phases. Older blocks were incorporated and fitted in. The tectonic structure is largely dependent on the nature of the basement rocks and the thickness of the young sediments. In the southern part of Middle Anatolia, three large massifs, i. g. the Menderes, Kırşehir, and Bitlis Massif, extend from west to east. Their cover is thin; even today it is almost horizontal and only displaced by faults. The Tuz Gölü Basin, too, even though it contains a sedimentary sequence of more than 3000 m (Turk. Gulf Oil Co., 1961), remained unfolded due to its crystalline basement. In the northern part of Middle Anatolia, in the area between the three massifs and the Pontides, structures are more varied. Close to the Aegean coast, between Çanakkale, Izmir, and Balıkesir, north-trending, mainly east-verging folds and thrusts occur (*Aygen*, 1956; *Radelli*, 1970; *Brinkmann* et al., 1970, 1972). Between Bursa

and Ankara, they are replaced by west-trending germanotype folds (*Abdüsselâmoğlu,* 1959; *Kalafatçıoğlu* and *Uysallı,* 1964; *Eroskay,* 1965; *Altınlı* et al., 1970b, 1972). East of Ankara, the Middle Anatolian ophiolite zone exhibits folds and upthrusts that initially were formed by southeast-verging and subsequently by south-verging compressional movements (*Norman,* 1973b). In the northeastern part of Middle Anatolia, a series of subsidence areas originated in Late Cretaceous time. The shallower basins of the Amasya-Erzincan-Erzurum zone were only subjected to faulting (*Irrlitz,* 1972). In the 5000 m-deep basin of Sivas, germanotype folds turning into regular ones are developed (*Kurtman* and *Akkuş,* 1971). In addition, there are some salt domes containing Oligocene salt (*C. E. Taşman,* 1937).

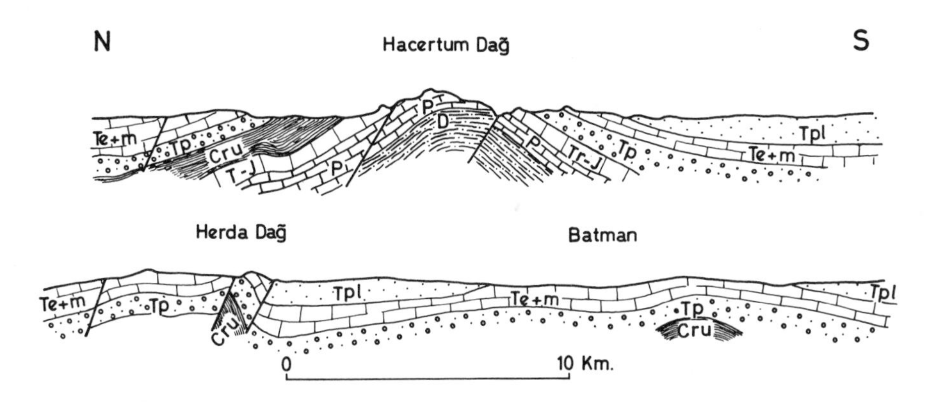

Fig. 41 Two sections across foreland folds in Southeast Anatolia (after *Tolun,* 1949)
D = Devonian, P = Permian, Tr-J = Triassic-Jurassic, Cru = Upper Cretaceous, Tp = Paleocene, Te + m = Eocene and Miocene

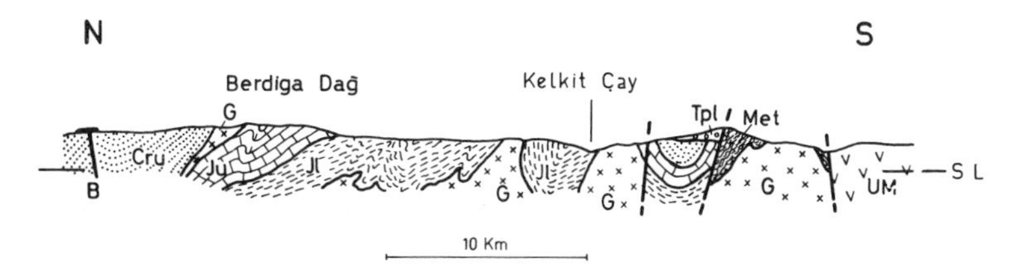

Fig. 42 Section from the southern border of the Outer Pontides across the Inner Pontides to the Middle Anatolian ophiolite zone (after *Nebert,* 1961a)
Met = Metamorphics, Jl = Liassic + Dogger, Ju = Malm + Lower Cretaceous, Cru = Upper Cretaceous flysch, Tpl = Pliocene, UM = Ultramafitite, G = Granite, B = Basalt

The young Alpidic tectonics of Turkey shows two distinctive features. First, the vergence is mostly to the south resulting in a north-dipping foliation (*Ketin,* 1947a: 260). There are some exceptions. The Pontides are north-vergent; several Middle Anatolian basins were overthrusted from both sides (*Ketin,* 1956; *Blumenthal,* 1952b, 1955: 146; *Kraus,* 1958). Second, the tectonic structures, except for Southeast Anatolia, do not extend on wider distances in strike. It is conceivable that this pattern was generated by tectonic forces which in the course of geological time changed in direction.

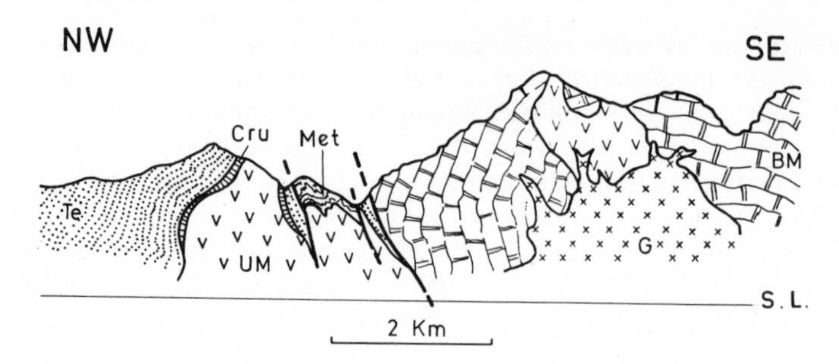

Fig. 43 Section across the Bolkar Dağ (after *Blumenthal,* 1955; *Demirtaslı* et al., 1973)
Met = Schist, BM = Bolkar Dağ marble (Permian to Cretaceous), Cru = Maestrichtian/Paleocene,
Te = Eocene flysch, UM = Ultramafitite, G = Granite

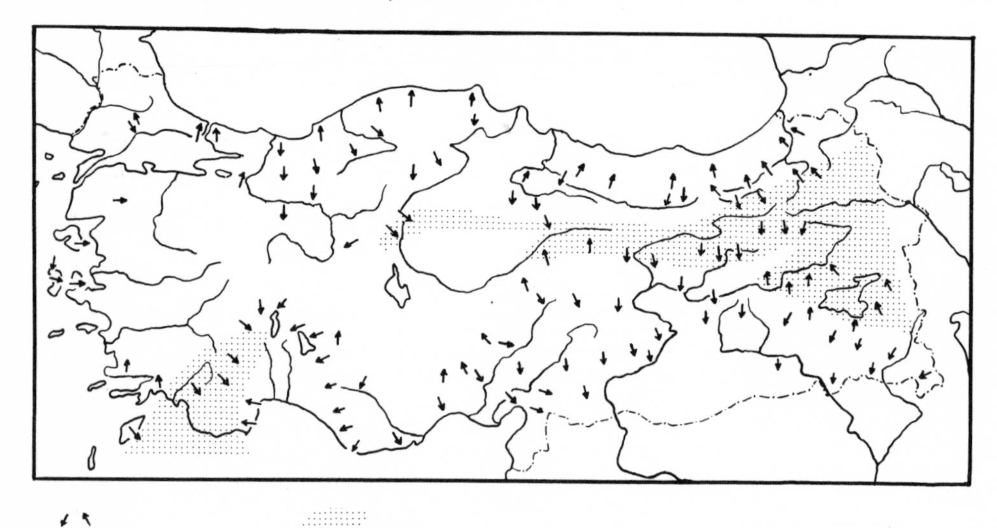

Fig. 44 Vergence of Alpidic structures and area of intense intra-Miocene tectogenesis
(after *Blumenthal,* 1952b; *Ketin,* 1956; *Kraus,* 1958, and others)

Late Alpidic Tectogeneses (Fig. 68)

The late Alpidic period represents the onset of a neotectonic epoch. New tectonic trends appear and dissect the older structures. Compression is replaced by extension. Normal faults are prevalent, thrusting is very rare (*Nebert,* 1958).

West Anatolia was fractured into blocks by west- to southwest-striking faults. At the coast, the horsts appear as peninsulas and the grabens as bays. In the interior, they form mountain chains and intermontane basins, i. e. ovas (*Salomon-Calvi,* 1936). The valley of the Büyük Menderes was already a rapidly subsiding basin during the Late Tertiary. In the early Pleistocene it turned into a graben with a vertical displacement of up to 1000 m (*Becker-Platen,* 1970: 164). The late Pliocene peneplain became dissected and displaced. In Thracia, its present elevation is 200 to 300 m, but in the Tekirdağ, it rises to 950 m (*Kopp* et al., 1969: 81). In the trough of the central part of the Sea of Marmara, it subsided at least 1350 m (p. 82), and in the Ulu Dağ it was raised to 1600 to 2000 m (*W. Penck,* 1918: 33). Similar observations were made near Ankara (*Erol,* 1961: 80, 1973). The height of the peneplains, therefore, cannot be used to determine their age; a fact which has to be remembered in connection with the dating of the old land surfaces in the Anatolian mountains (*Zwittkovits,* 1966).

In the Taurus, the young faults are mostly north-trending. They have mainly formed the highlands, ovas and lake basins of this area (*De Planhol,* 1952; *Becker-Platen,* 1970: 212; *Bering,* 1971: 129).

In Hatay, the fault lines of the Jordan system cut the Southeast Anatolian tectogene and lead to northeast-trending horsts and grabens. Basaltic eruptions characterize the edges of the grabens and give a clue to the age of the block structures. They are of Quaternary age but go back as far as the Eocene (*Schwan,* 1972: 150).

East Anatolia was less affected by the Wallachian tectogenesis. Here, the Upper Miocene-Pliocene strata are mostly undisturbed, except for the basins of Amasya-Erzurum (*Irrlitz,* 1972; p. 126).

Alpidic Regional Metamorphism

As stated on p. 12, part of the crystalline schists of Turkey formed during the Alpidic era. This metamorphism falls into two facies series which also differ in their geologic appearance.

Facies Series of the Barrow Type (Fig. 45)

The metamorphites of this series, in conjunction with old crystalline rocks, form a series of massifs which already are described (p. 5):

1. The northeastern part of the *Istranca Massif* and the adjoining Sakar-Strandca zone in southern Bulgaria consist of a thick, mostly low-metamorphic sequence of Triassic and Jurassic age. Foliation and recumbent folding of northeasterly vergence is developed. It seems not necessary to postulate nappe structures with windows (*Tollmann,* 1965). Probably this mountain range is the western extension of the Inner Pontides (p. 47). However, the folding proceeded later than in North Anatolia. It may be associated with Early Cretaceous metamorphism before Cenomanian time (*Pamir* and *Baykal,* 1947: 40). The Istranca Massif remained an uplift area for long periods, with the exception of partial flooding during the Late Cretaceous and Early Tertiary.

   Mesozoic metamorphites are exposed in the Greek part of Thracia along the western edge of the Thracian Tertiary Basin. It is possible that the basin formed between the two diverging zones of young tectogenesis and metamorphism (*Ivanov* and *Kopp,* 1969b).

2. The *Menderes Massif* was formed mostly by the Alpidic metamorphism. This process took place in several stages, the last one of Early Tertiary age (p. 9). The present shape of the massif emerged since the Cretaceous. It is almost entirely surrounded by Alpidic tectogenes, in the south by the Taurus, in the northwest and north by the Middle Anatolian ophiolite zone. At the end of the Tertiary, the massif subsided and became a block faulted area (*Arpat* and *Bingöl,* 1969; *Becker-Platen,* 1970).

   The western extension of the Menderes Massif, the Cyclade Massif, has been submerged in the Aegean Sea. Recent investigations have shown the existence of metamorphic phases of similar age (*Brinkmann,* 1971a: 896). It is unclear how far the Menderes Massif extends to the northeast. Radiometric dating of phyllite west of Ankara gave an age of 156 m. y. (*Çoğulu* and *Krummenacher,* 1967).

3. In the *Kırşehir Massif* a young metamorphic phase could not be ascertained up to now (p. 10).

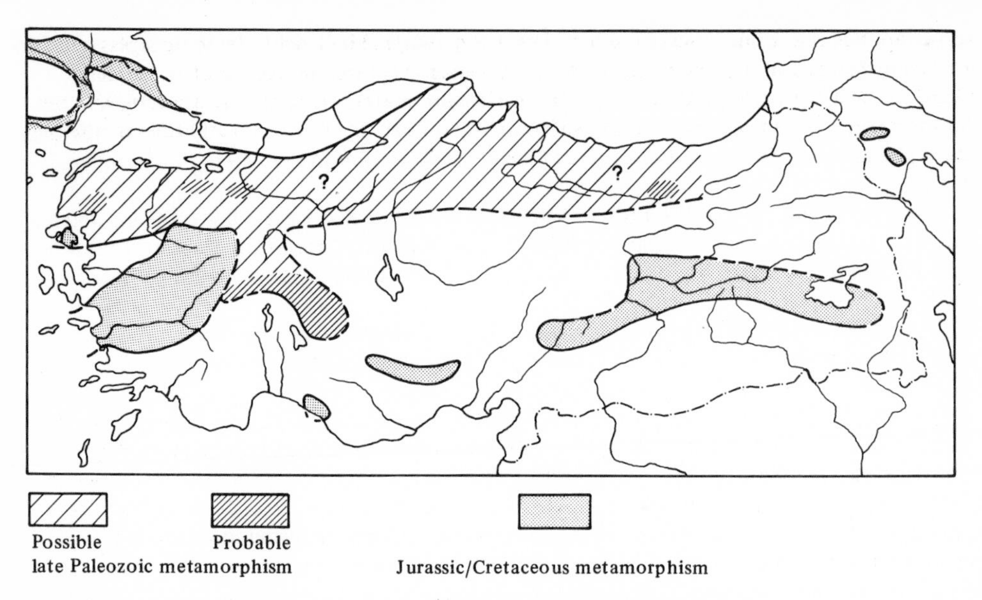

Fig. 45 Distribution of metamorphic events of Variscan and Alpidic age in Turkey

4. The summits of the *Bitlis Massif* are composed of crystalline schists and marbles of Permian age (p. 10). The Alpidic metamorphism, however, has a larger areal extent. The Bitlis Mountains are surrounded by a wide aureole where the Permian limestone and its underlying strata are recrystallized, too. It gradually disappears near Tunceli in the north and Malatya in the west (*Blumenthal,* 1944a: 123). In the south it is sharply limited by the Bitlis overthrust – an indication of the width of crustal shortening. Toward the southeast, the aureole continues into Iran and runs parallel to the Zagros line forming the Sananday-Sirjan zone (*Stöcklin,* 1968a: 1247). Here, metamorphism is of Lower Cretaceous age (*Braud,* 1973).

5. In the Massif of Alanya-Anamur, it is uncertain whether the last metamorphism is Mesozoic or Paleozoic (p. 12).

6. The metamorphic aureole which extends from Konya Ereğlisi into the Bolkar Dağ exhibits marbles of Permian to Late Cretaceous age (*Blumenthal,* 1955: 141; *Brinkmann,* 1971a: 893; *Demirtaşlı* et al., 1973).

Lawsonite-Glaucophane Facies Series (Fig. 31)

In Anatolia, the rocks of this series appear as geological bodies of small dimensions, which are evidently associated with the ultramafitite massifs (*Van Der Kaaden,* 1966, 1969). In the ophiolite zones of Middle and Southwest Anatolia they belong to the lawsonite-albite or lawsonite-glaucophane facies. In Southeast Anatolia metamorphism rarely exceeds the pumpellyite-quartz grade (*Hall* and *Mason,* 1972: 396; *Pişkin,* 1972: 119; *Yazgan,* 1972: 207; *Aslaner,* 1973: 68). Detailed studies indicate that the metamorphites emerge from ophiolite sequences. They rest on a base of ultramafitite and are frequently covered by unmetamorphosed ophiolite strata (p. 86). The metamorphism has left the fabric of the parent rocks remarkably untouched; schistosity is missing. The major problem in this case is where to get the high pressure of 5.5 to 8 kb, which is needed according to experiments.

A metamorphism by burial is hardly plausible in Anatolia. The sedimentary overburden may have reached thicknesses of up to 1 to 3 km; on geological grounds, higher thicknesses are unlikely. The same applies to the idea of tectonic overburden. Thrust plains or remains of nappes are missing. Under such circumstances it is questionable how far tectonic stress and internal gas pressure can provide conditions necessary for the lawsonite-glaucophane facies (*Winkler,* 1967: 170). Structural relationships in Anatolia suggest that the migration of volatiles from the ultramafitite into its ophiolitic roof has played a part in the metamorphism (*Brothers,* 1970).

Chapter 18 Recent Crustal Movements

Epirogenic Processes

In Turkey, the buildings of ancient Greek and Roman time give a unique opportunity to study the relation between land and sea during the last millenia. Observations in the western and southwestern parts of Anatolia show mostly subsidences, rarely uplifts by about 1 to 2 m since that time (*Tietze,* 1885: 367; *Philippson,* 1920: Tab. 3; *Hafemann,* 1960; *Schäfer* und *Schläger,* 1962; *Eisma,* 1962; *Flemming,* 1972). These phenomena can best be explained by tectonic rather than eustatic causes.

In the interior of Turkey repeated geodetic measurements are not available. In the adjacent areas of Russia such data exist. For instance, Armenia is rising by 2 mm/yr, while the Black Sea coast is subsiding by the same amount (*Dumitrashko* and *Lilienberg,* 1967).

Earthquakes (Fig. 46–48)

Turkey ist one of the most frequented earthquake centers of the world. In the last 40 years alone, 50000 people died and hundreds of thousands of buildings were destroyed. It is true, however, that the customary way of house construction plays a fateful role (*Ergin* et al., 1967, 1971; *Karnik,* 1971; *Ilhan,* 1971b; *Allen,* 1975). Six zones of particular earthquake activities are known:

1. The *North Anatolian earthquake zone* (An. 1973) is the most intensive one; magnitudes of up to M = 8.1 have been recorded. It runs about parallel to the Black Sea coast throughout Anatolia (*Ketin,* 1969). In the west, in the area of the Sea of Marmara it splits (*Kopp* et al., 1969: 104). In the east, it divides into two branches which continue toward the Van Gölü and Armenia. The zone is characterized by a vertical fault plane which is repeatedly displaced en echelon. In the landscape it occurs as a chain of depressions, which, in part, are occupied by river valleys, lakes, or bays. It is accompanied by thermal springs, locally also by minor basaltic eruptions. The basement rocks are mylonitized in a belt several kilometers wide.

   Along this zone, earthquakes occur in intervals of tens of years and the epicenters sometimes shift regularly (*Karnik,* 1971: 191). The isoseismals run parallel to the earthquake zone. During quakes, faulting takes place. In the horizontal, the displacement reaches up to 4.5 m, always in a right-lateral sense, and in the vertical 1.5 m whereby normally the northern block subsides. The North Anatolian seismic line constitutes a right-lateral strike-slip fault (*Pavoni,* 1961); this assumption is confirmed by investigations on focal mechanism. The movements continue also during times of quiescence; constant creep results in a displacement of several cm/yr (*McKenzie,* 1972: 176).

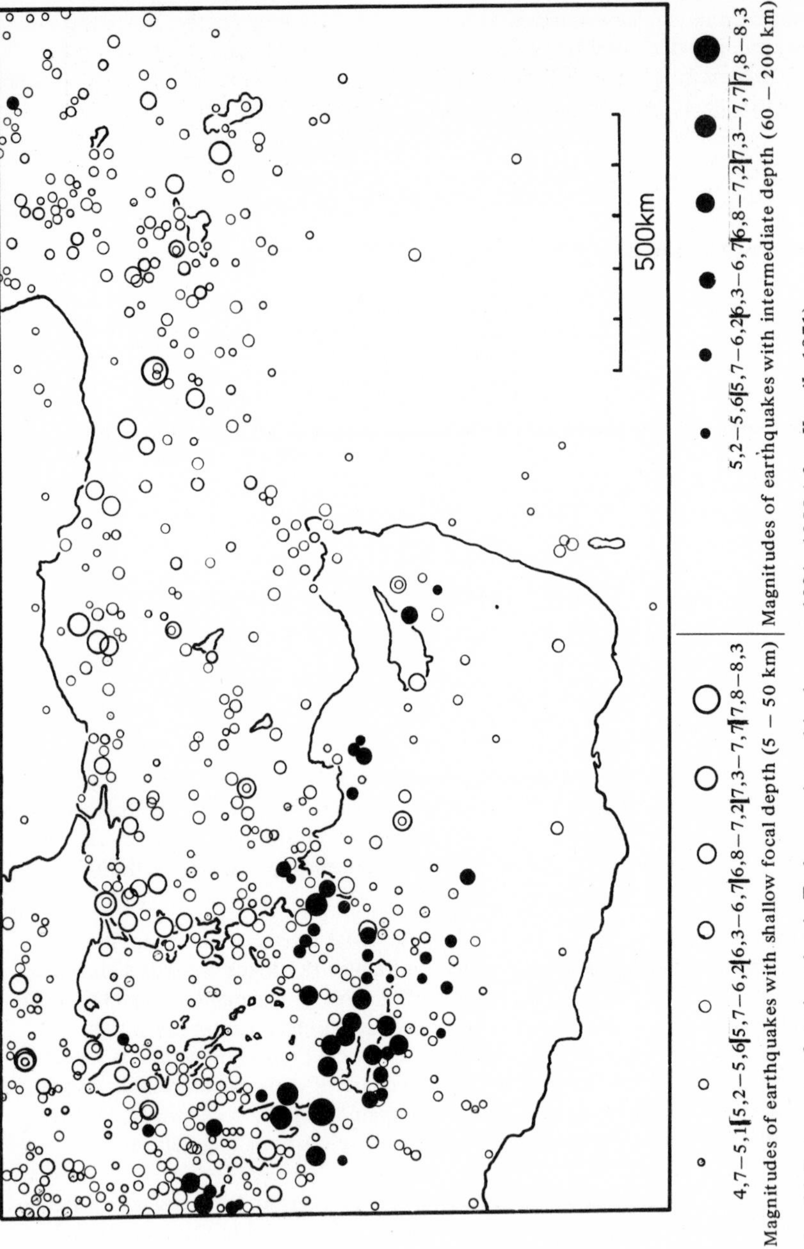

Fig. 46 Epicenters of earthquakes in Turkey, registered in the years 1901–1955 (after *Karnik*, 1971)

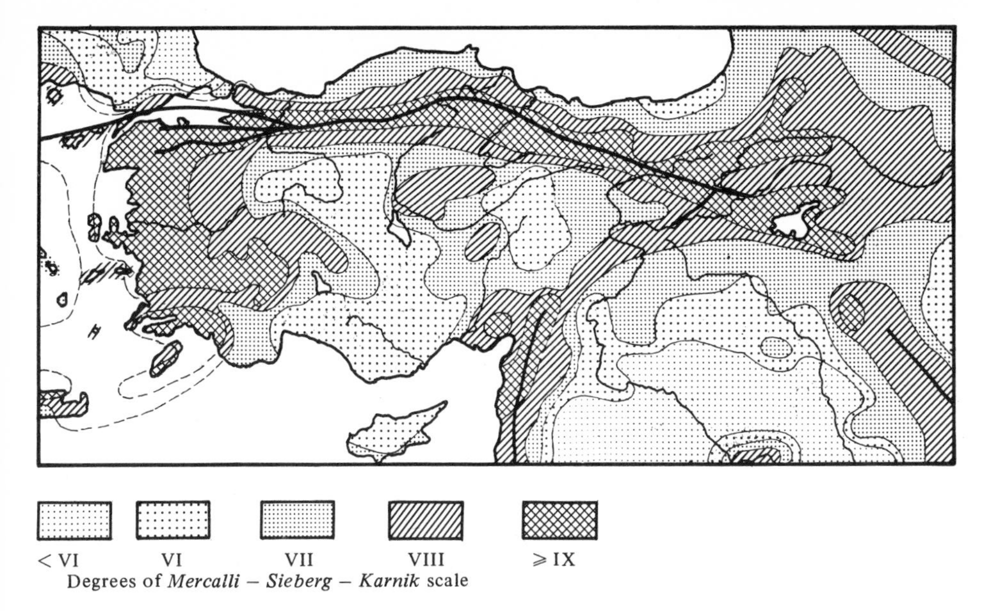

Fig. 47 The average maximal intensity of earthquakes in Turkey, and the main earthquake lines (after *Ketin,* 1969; *Karnik,* 1971, and others)

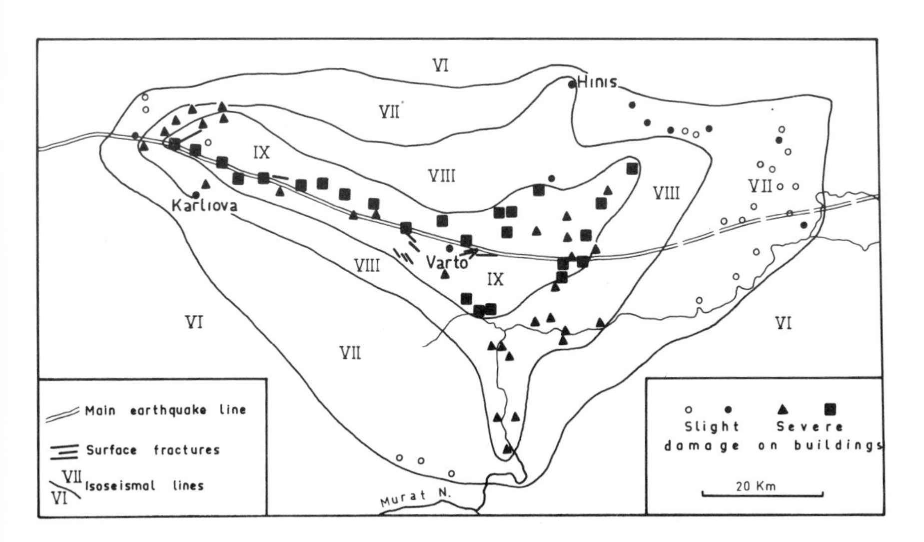

Fig. 48 Isoseismal map of the earthquake on 16.8.1966 near Varto, eastern Anatolia (after *Ketin,* 1969)

2. In the continuation of the North Anatolian line, an earthquake zone emerges southeast of the Van Gölü which traverses the entire Iran. It follows the *Zagros line,* which seismologically appears partly as an upthrust, and partly as a right-lateral wrench fault (*Canitez,* 1969).

3. The *East Anatolian* or *Bingöl earthquake zone* branches from the North Anatolian zone northwest of the Van Gölü in a 60° angle; it can be traced via Bingöl to the Hazar Gölü south of Elaziğ. It represents a left-lateral wrench fault in which a displacement by 15 to 25 km may have occurred since the end of the Tertiary period (*Arpat* and *Şaroğlu,* 1972; *Seyman* and *Aydın,* 1972).

4. The *earthquake zone of Hatay* is also associated with a left-lateral strike-slip fault. Toward the north it disappears near Maras, but there may be a connection with the East Anatolian zone. Toward the south, it continues via the Jordan line into the rift system of the Red Sea. Holocene horizontal displacements are known in Israel. In Turkey, only one young subsidence has been observed in Hatay (*Stark,* 1956).

5. *West Anatolia* and the adjoining parts of the *Aegean Sea* constitute an area with frequent earthquakes. Here the epicenters follow lines which coincide with the faults of the Wallachian phase (*Pınar,* 1949). In West Anatolia, the seismic lines are west-trending, and in Southwest Anatolia they turn into a southwestern to southern direction. Seismically, the earthquakes are partly normal faults and partly wrench faults with movement planes of a mean strike of N50°E (*Ambraseys* and *Tchalenko,* 1972).

6. The earthquake zone of the western Taurus is discussed on p. 111.

The seismic centers of the majority of earthquakes in Turkey have a focal depth of 10 to 30 km. Some earthquakes of an intermediate focal depth occur in the eastern part of the North Anatolian earthquake zone and in the western Taurus. Deep-focus events are not known so far.

All seismic lines cross the tectonic structures of Turkey. On the whole, they are thus younger than the last major phases of the Alpidic tectogenesis. But locally they follow older patterns. In its middle part, north of Ankara, the North Anatolian seismic line runs parallel to the Middle Anatolian ophiolite zone (*Nowack,* 1928: 311; *Brinkmann,* 1968: 118). In southern Thracia, faults of Eocene age roughly coincide with the present seismic line (*Kopp* et al., 1969: 100). The chain of Tertiary basins in the northeastern part of Middle Anatolia represents a zone of subsidence which collapsed along a seismic line from the late Miocene to the Quaternary period (p. 93). The basin of Çeltek north of Amasya may have formed in a similar way and as early as Paleogene (p. 66). The history of the seismic lines in West and Southwest Anatolia is shorter. They follow neotectonic fracture zones, but they may, at times, have Late Tertiary precursors, as shown by the Büyük Menderes graben (p. 94). Even though part of the seismic lines may have antecedents, their activity did not start until fairly recent geologic time (*Ketin,* 1968: 66). Therefore, the present amount of displacement along the North Anatolian fault should not be extrapolated into Mesozoic time. Rivers have been diverted by not more than 1/2 to 1 km along the line. The two parts of the Tekirdağ anticline appear to have been displaced by 30 km since the Savic phase (*Kopp* et al., 1969: 73). Distances greater than that are not known so far (*Ketin,* 1969: 18).

Chapter 19 Magmatism

Magmatic Rocks and Mineral Deposits (Fig. 30, 49–51)

Observations on the magmatic rocks of Turkey are scattered in numerous geologic and economic publications. Summaries on the magmatic events have been prepared by *Angel* (1931) and *Ketin* (1961) for the entire country and by *Van Wijkerslooth* (1944) and *Westerveld* (1957) for the Cenozoic volcanism. *Borchert, Egeran, Gysin, Kovenko, Maucher, W. E. Petrascheck, Pollak, Van Wijkerslooth* and *Van Der Kaaden* are the principal investigators of the Turkish magmatic mineral deposits. These are enumerated by *Ryan* (1960), described in a series of monographs (Pub. MTA, since 1964), and shown on a map (*A. Gümüş,* 1970, 1972).

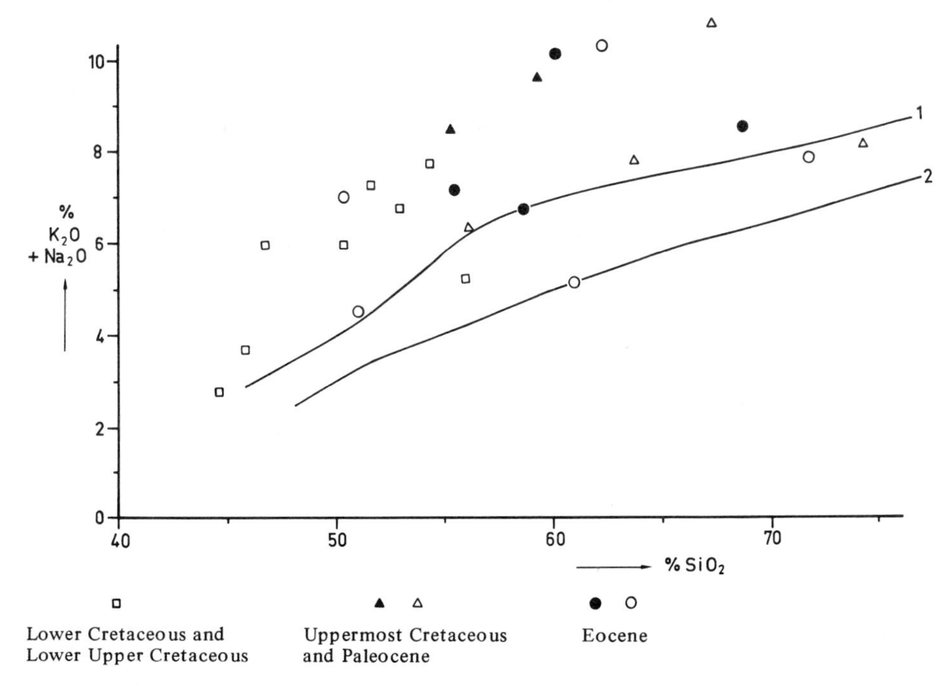

Fig. 49 Chemical composition of the Cretaceous/Paleogene plutonics (full signs), and volcanics (open signs) in the eastern part of the Outer Pontides near Ordu (after *Gedikoğlu,* 1970, and *Kilinc,* 1971). Lines 1 + 2 see Fig. 30

The Cryptozoic and Paleozoic

The crystalline massifs contain magmatites which may go back as far as Cryptozoic time. The magma was evidently poor in volatile components. Pegmatite veins are rare, and magmatic ore deposits are missing altogether.

The earlier Paleozoic era was a time of relative magmatic quiescence. The Cambrian to Devonian strata only contain occasional intercalations of basic volcanite or tuff. During the later Paleozoic, these rocks become more frequent, such as in the lower Carboniferous strata at Istanbul, in the Orhanlar Formation of West Anatolia, or in the Permian of the western Taurus. In addition, there are several granitic plutons ranging in age between 338 and 272 m. y.

Mineral deposits of definite Paleozoic age are rare. The wolframite ore of the Ulu Dağ may belong to this period (*Öztunalı*, 1973: 10). An uraninite vein in the gneiss core of the Menderes Massif was dated at 268 m. y. (p. 9). The hematite lenses in the Devonian diabase tuffs on the Antitaurus can be interpreted as exhalative-sedimentary deposits (*Vaché*, 1964: 94). The lead-zinc ores which often occur in the Permo-Carboniferous limestones, however, are probably very much younger than the adjacent rocks.

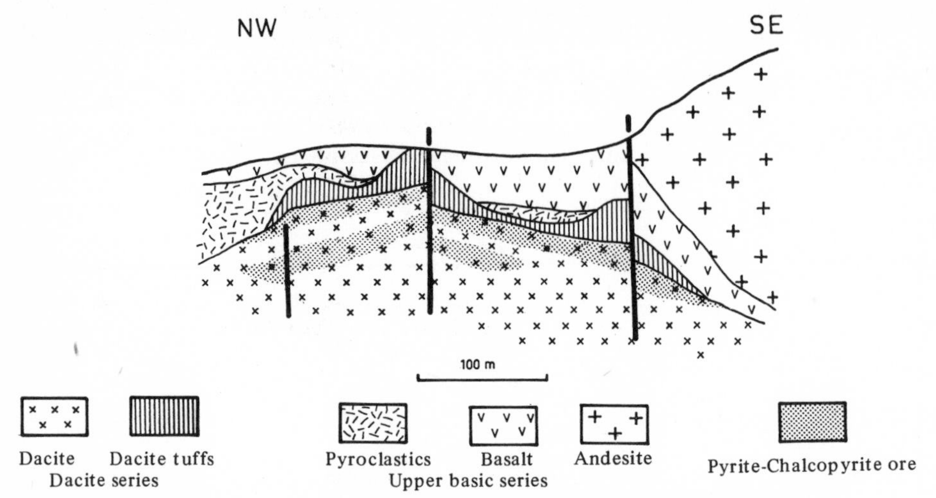

Fig. 50 Section across the copper ore deposit Lahanos near Giresun (after *Pollak*, 1961)

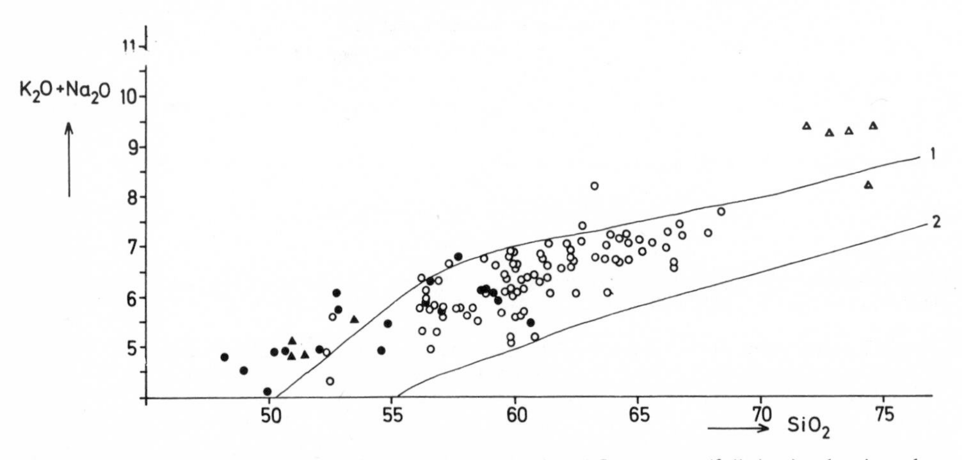

Fig. 51 Chemical composition of the Neogene (open signs) and Quaternary (full signs) volcanic rocks between Konya and Kayseri (after *Jung* and *Keller*, 1973). Lines 1 + 2 see Fig. 30

The Mesozoic

Beginning with the Triassic period, the magmatic activity increased. It was not distributed evenly across Turkey but was concentrated in several geosynclines within the Tethys. Each geosynclinal type was characterized by a different association.

During Late Triassic time, the first ophiolite eugeosynclines formed in the western Taurus, at Ankara, in Southeast Anatolia, and possibly also in other areas (p. 44). They reached their greatest extension during the later Mesozoic period, and several of them existed until early Tertiary time. All of them were associated with a strong, mostly alkaline-basaltic volcanism, which occasionally furnished andesitic-trachytic differentiates (p. 84). Ore deposits as a result of the basaltic eruptions are not known. The emplacement of the ultramafitites, however, was of major significance to the mining industry in Turkey. It carried the chromite ore to the surface.

The subsidence of the Pontian geosynclines, which represent normal eugeosynclines, started in Liassic time. It was accompanied by eruptions of intermediate composition, especially in the eastern part of North Anatolia. In the Inner Pontian geosyncline, spilitized basalt and keratophyre are prevalent in the Liassic-Dogger (*Nebert,* 1964b: 44). In the Outer Pontian geosyncline, a cycle took place from the Malm to the Late Cretaceous epoch that led from basalt to andesite and trachyandesite, and finally to dacite and rhyolite (*Maucher* et al., 1962: 12; *Gedikoğlu,* 1970: 19). During that period the eruptions spread in the geosyncline. The basic eruptions were essentially restricted to the eastern part. The intermediate and acidic eruptions reached as far as Eregli to the west (*Blumenthal,* 1950: 67; *Fratschner* and *Van Der Kaaden,* 1953). At Bolu and Kocaeli they advanced across the southern rim of the geosyncline (p. 55). Manganese ore lenses in Upper Cretaceous strata may be due to exhalation (*Chazan,* 1947; *Maucher* et al., 1962: 69). The tectogenesis of the Inner Pontides was accompanied by granodioritic-dioritic plutonism, especially in the area between Inebolu and the upper Kelkit valley. According to *Kovenko* (1944), the hydrothermal pyrite-chalcopyrite deposit of Küre south of Inebolu was formed at this time. But granitic magma also erupted in the northern part of the Menderes Massif where it caused minor lead-zinc-copper mineralization at the contact (*Dora,* 1969).

During the Jurassic and Early Cretaceous period, the western Taurus was a miogeosyncline. Several basalt sheets which are intercalated in Malm strata, represent the only eruptions during that time (*Poisson,* 1968; *Haude,* 1972: 417; *Desprairies* and *Gutnic,* 1972). Apparently the ophiolitic volcanism was inactive and did not revive until the middle and late Cretaceous (p. 57).

The turn of the Cretaceous was of importance for magmatic processes. Several granitic massifs of the western and central part of Anatolia are radiometrically dated at 74–68 m. y. (p. 55). Rocks with a similar age appear to exist also in East Anatolia (*Nebert,* 1964b: 49). This suggests that during the latest Cretaceous and the earliest Tertiary epoch, granitic magmas erupted over a wide area.

The Paleogene

Starting with the Tertiary period, a new cycle of eruptions took place in the Outer Pontian geosyncline. The extrusive rocks mainly consisted of basalt but also of rhyodacite and trachyte. This cycle lasted until the end of the Eocene epoch (*Maucher* et al., 1962: 12; *Gedikoğlu,* 1970: 52). The Pyrenean folding was accompanied by intrusions of granodiorite and quartz-syenite. Probably shallow laccoliths formed, because the plutonite frequently shows a porphyritic marginal facies or develops into rhyolite (*Maucher* et al., 1962: 81; *Çoğulu,* 1970: 39; *Kilinç,* 1971: 88). Massifs of that age are particularly widespread in the eastern Pontides, but they are also present in their western part (*Blumenthal,* 1950: 47).

Controversy exists concerning the age of the pyrite-chalcopyrite ores which, in the eastern Pontides, in many places are present in the form of vein fillings, beds, and breccia zones. *Maucher* et al. (1962: 13) regard them mainly as exhalative-sedimentary formations of Cretaceous age. According to *Pollak* (1968), *Bernard* (1970), and *Pejatović* (1971: 15), the ores are of hydrothermal origin, except for a few contact-pneumatolitic occurrences. The minor part is supposed to belong to the Late Cretaceous (*Gedikoğlu,* 1970: 99), and the major part to the Eocene. In the ore districts of Transcaucasia similar conclusions have been attained.

Middle Anatolia, too, has several granitic massifs and ore deposits of probably early Tertiary age (*Ketin,* 1959c: 166; *Norman,* 1972: 242). At Divriği east of Sivas, a hematite deposit formed in the contact zone between syenite and Cretaceous limestone (*Kovenko,* 1939; *Klemm,* 1960). According to *Van Wijkerslooth* (1954), the copper sulfide ore of Ergani Maden south of Elaziğ represents a post-tectonic impregnation of ophiolitic rocks in the Southeast Anatolian zone. The K/Ar age of 31.5 ± 0.8 m. y. of an adularia from the mining area coincides with these dates (*Griffith* et al., 1972: 714). The galena-sphalerite deposits of Keban (*Kumbasar,* 1964; *Kineş,* 1969), Akdağmadeni (*Vaché,* 1963; *Pollak,* 1958), Balya (*Gjelsvik,* 1962), and the Antitaurus (*Petrascheck,* 1967) are considered to be of similar age.

Following the Pyrenean intrusive phase, magmatic processes in the Outer Pontides ceased to exist until the Pliocene epoch. In contrast, volcanic activity started to appear in the northern parts of Middle Anatolia. The strata in the Tertiary basins (p. 66), particularly between Yozgat, Sivas, and the upper Kelkit valley, contain lavas and tuffs of andesite and basalt. The eruptions reached their peak during the Middle to Late Eocene epoch (*Kurtman,* 1973a: 20).

In contrast to the Pontides, the tectogenesis of the western Taurus took place without accompanying magmatic symptoms.

The early Tertiary volcanic rocks of Thracia fall between latite-andesite and rhyolite. They are mostly restricted to the slope of the Rhodope Massif toward the Thracian Basin, but they also advance to the Biga Peninsula across the Straits (*Ivanov* and *Kopp,* 1969a).

The Neogene-Quaternary

Several early Miocene granite masses in the western part of Middle Anatolia are the youngest visible plutonic rocks in Turkey (p. 72). Also some subvolcanoes have been exposed since late Tertiary time (*Nebert,* 1961b). But otherwise, only surficial volcanic formations are known. Those formations underwent a remarkable change about the middle of the Tertiary period. At the same rate as the Alpidic tectogenesis progressed, the volcanism migrated from the geosynclines to the surrounding areas. The eruptions changed from a linear to an areal distribution. The chemical composition did not alter very much at first. As in previous periods, intermediate calc-alkalic rocks were predominant as late as the early Miocene epoch, with quartz-latite („andesite“) in first place, followed by dacite and basalt (*Dora,* 1964; *Savaşcın,* 1974). During the Miocene and particularly during the Pliocene, rhyolite, rhyodacite, and obsidian on the one hand, and alkali-basalt on the other appeared increasingly (*Maucher* et al., 1962: 12, 42, 86; *Özsayar,* 1971: 37; *Jung* and *Keller,* 1972; *Ayranci* and *Weibel,* 1973). At the Aegean coast this change occurred 16 to 12 m. y. ago (*Borsi* et al., 1973). On the whole, the Anatolian volcanism reached its peak during the late Miocene-early Pliocene. Since that time, it has decreased in intensity.

On a regional scale, the Upper Pliocene alkali-basalts of Giresun-Trabzon represent the only late Tertiary volcanites of Turkey that settled on an Alpidic tectogene. The other young mountain ranges are remarkably free of Neogene-Quaternary volcanic formations. Older cratonic areas, however, such as West Anatolia, the so-called Paphlagonian Massif north of Ankara, the Kırşehir Massif (*Pasquaré,* 1968), and the eastern part of Middle Anatolia (*Pasquaré,* 1971), were covered by thick sequences of lava and tuff, often ignimbrite. In Southeast Anatolia, magmatism was of a special kind. As in Syria and Jordania, Quaternary flood basalts are the only young eruptions.

Here and there, the plutons and subvolcanoes of Middle Anatolia are accompanied by minor oxidic and sulfidic mineralization (*Dora,* 1964; *Jacobson* and *Türet,* 1970; *Leo,* 1972). Some afterevents of the early Tertiary copper ores appear in the Pontides (*Pollak,* 1968: 92; *Pejatović,* 1971: 18). The boron and mercury deposits are of economic importance. The borate is regarded as a precipitate from volcanic exhalations in an arid area (*Özpeker,* 1969; *Brown* and *Jones,* 1971; *Baysal,* 1973; *Demircioğlu,* 1973). The cinnabarite impregnations are very young and limited to the Wallachian faults (*Lehnert-Thiel,* 1969; *Arpat* and *Bingöl,* 1969: 3).

Thermal Springs (Fig. 52)

At present no volcanoes are active in Turkey (p. 81). But their aftereffects can still be recognized by some fumaroles and over 300 thermal springs (*Caglar,* 1947/61; *C. Erentöz* and *Ternek,* 1968; *Kurtman,* 1973b). This huge number is due to especially favorable conditions. First of all, the geothermal gradient in parts of the country may be steeper than normal, although no data exist. Second, the Wallachian tectogenesis, which was accompanied by expansion, created numerous deep fissures. Rainwater sinks down and rises again as a hydrothermal solution, enriched with dissolved material leached from the adjacent rocks. Close to the coast, sea water rises in the same manner (*Brinkmann* and *Kühn,* 1973). The thermal springs are often situated on seismic lines (*Pınar,* 1949). West Anatolia consists of a block-faulted area of Quaternary age with frequent earthquakes; therefore, it has the hottest springs (see Fig. 47).

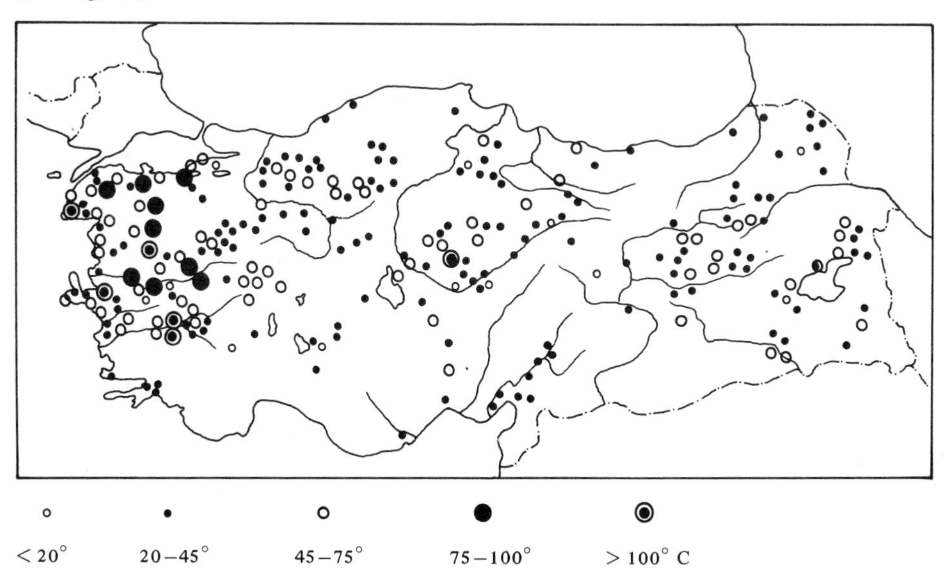

< 20° 20–45° 45–75° 75–100° > 100° C

Fig. 52 Distribution of thermal springs in Turkey (after *C. Erentöz* and *Ternek,* 1968)

Summary (Fig. 53)

For most of its geologic history, the magmas of Turkey belonged to the calc-alkalic assemblage. It was not until the end of the Tertiary period that a preference for the alkalic assemblage. arose. Such a development is characteristic for mobile areas subjected to a gradual consolidation.

The plutons show only minor temporal and regional differences. Granodiorite is constantly prevalent. Starting with the later Cretaceous epoch, monzonites and syenites appear in addition to granodiorites. The rise of magma was roughly a time-equivalent to the tectogenic events. According to radiometric dates, the following groups can be distinguished:

305–290	270	160	74–68	60–55	40–35	25–22 m. y.

which correspond approximately with the

middle Variscan	late Variscan	late Kimmerian	Ressen	Laramide	Pyrenian	Savic-Styrian

tectogeneses.

During the Paleozoic, Mesozoic, and early Tertiary periods, volcanism was limited to the geosynclines within the Tethys. The majority of eruptions were andesites and basalts. At the same rate as the geosynclines were closed by folding, volcanism appeared in the areas between the young tectogenes. Even during Neogene and Quaternary time, intermediate lavas of the calc-alkalic assemblage made up most of the eruptions. Since the Pliocene epoch, however, rhyolitic and basaltic alkalic rocks appeared in increasing quantity – almost simultaneously with the neotectonic blockfaulting. A temporal migration from west to east in the volcanic fiels of Afyon-Konya (at the turn of the Miocene), Nevşehir-Ürgüp (early Pliocene), and Erciyes (Quaternary) can be recognized. So far, a connection of the volcanic phenomena and subduction zones has not been shown.

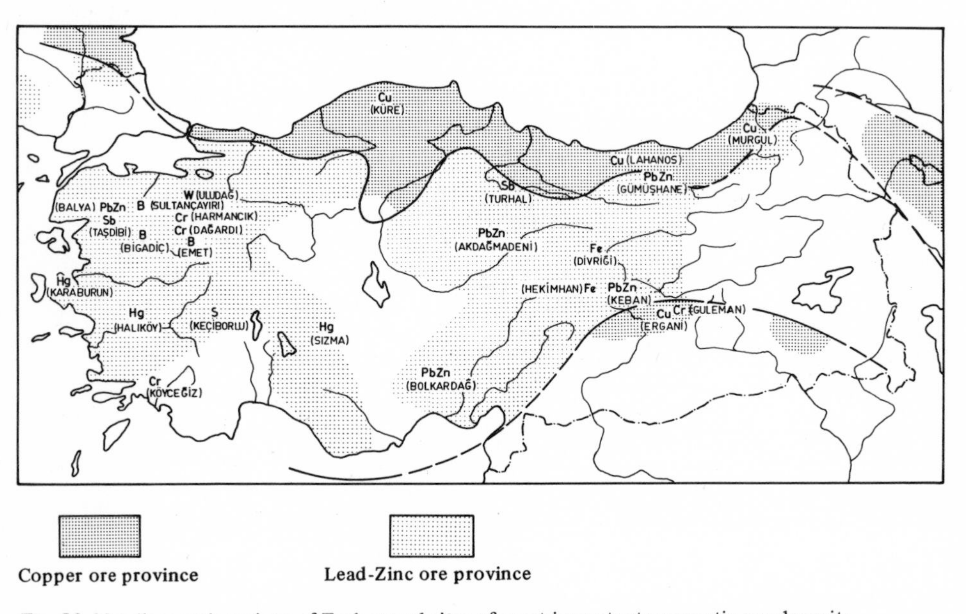

Fig. 53 Metallogenetic regions of Turkey and sites of most important magmatic ore deposits (after *Egeran,* 1946; *Petrascheck,* 1954/55; *A. Gümüş,* 1970)

The old crystalline basement is remarkably devoid of mineral deposits. Also Paleozoic and early Mesozoic ore occurrences are rare, with the exception of the chromite deposits. The major mineralization started in Late Cretaceous time and reached its peak during the early Tertiary, particularly immediately after the Pyrenean tectogenic and intrusive phase. At that time the separation into several metallogenetic zones occurred (*Egeran,* 1946; *Petrascheck,* 1954/55, 1964). The Pontian copper ore belt is situated in the north and extends to Bulgaria and Yugoslavia in the west and to Afghanistan in the east. Middle and South Anatolia belong to a lead-zinc ore province which also reaches into the neighboring countries. A second copper ore belt appears to exist in Southeast Anatolia.

On the whole, the magmatic phenomena in Turkey increased in intensity and extent from the Paleozoic to the Cenozoic era. During that time, it appears that the basement has been subjected to increased heating. Under these conditions, anatectic processes could rise in the crust. Maybe the widespread distribution of andesitic rocks in Turkey, which, according to *Jung* and *Keller* (1972: 511), *Günther* and *Pichler* (1973: 411), and *Borsi* et al. (1973: 491), are partly derived from crustal matter, can be explained that way.

Part Three The Geotectonic Position of Turkey

Chapter 20 Seas Surrounding Turkey

The seas of the eastern part of the Mediterranean region have recently been described in monographs: the Black Sea by *Degens* and *Ross* (1974; see also *Laking,* 1974) the Mediterranean by *Ryan* et al. (1970), *Stanley* (1972, with map), and *Ryan, Hsü* et al. (1973a), the Aegean Sea by *Maley* and *Johnson* (1971).

The Black Sea (Fig. 54)

The shelf of the northern coast of Anatolia is narrow. At a depth of 200 m, the shelf slopes to the central basin which is occupied by a deep-sea plain at 2100 m. According to seismic data (*Neprochnov* et al., 1974), the deep-sea basement is composed as follows:

1. The uppermost layer is 8 to 14 km thick with v_p = 1.8–4.5 km/sec, apparently a young, unsolidified, sedimentary sequence. If the present rate of sedimentation of 10 cm/1000 yr in the Black Sea is applied (*Ross* and *Degens,* 1974: 191), its sedimentation should have started 100 m. y. ago.

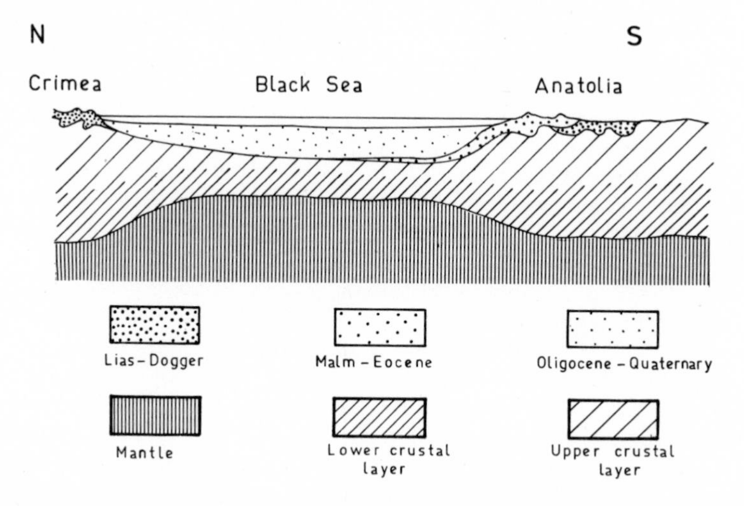

Fig. 54 Section across the Black Sea from Crimea Peninsula to North Anatolia

2. It is followed by a 5 to 10 km-thick layer with v_p = 6.6–7.0 km/sec, interpreted as the basaltic layer. The upper, granitic layer is supposedly missing beneath the deep-sea bottom and only present in the marginal part of the sea with v_p = 5.8–6.3 km/sec. *Rezanov* and *Chamo* (1969), however, argue that metamorphites could be present, because at a depth of 15 to 20 km they give the same wave velocity as basalt. Also the magnetic isanomals do not indicate principal differences between the center and the margin of the Black Sea. They run uniformly northwest-southeast throughout the entire area (Anom., 1965; *Ross* et al., 1974: 27;

Neprochnov et al., 1974: 45). The same direction prevails on the adjoining mainland in the north and coincides with the strike of the crystalline schist of the Scythian platform. These observations suggest a granitic layer in a reduced thickness to be present beneath the central portion of the Black Sea.

3. The mantle shows velocities of v_p = 8.0–8.2 km/sec. The Moho beneath the neighboring mainland is located at a depth of 30 to 50 km, and rises to 20 km in the center of the Black Sea. There is a corresponding increase in the Bouguer anomaly.

The structure of the crust is explained by the geologic history of the Black Sea region. There are several reasons to believe that a Pontian land has formerly occupied the area of the Black Sea (p. 17). The Black Sea region, related to Anatolia, took a marginal position to the Tethys from Paleozoic to Mesozoic times. First, the sections along the northern coast of Anatolia show more gaps than the ones in the interior. In the western part of the coast, Permian, Triassic, and Jurassic strata are largely missing, and in the eastern part, Paleozoic and Triassic strata are absent. Second, several systems, mainly the Ordovician, Upper Carboniferous, Permian, and Liassic-Dogger beds, which in the interior are marine, have along the Black Sea coast a continental, frequently coarsely clastic facies. Third, to the west, the basement of the Black Sea continues into the Moesian platform, which underlies the Bulgarian-Romanian lowland, and to the east, into the Transcaucasian Massif supposed to form the basement of the Rion-Kura depression. Both blocks continued to rise for long periods of time before they subsided in the later Mesozoic.

From the Late Cretaceous onward, however, there are no longer indications for a Pontian land (p. 59). Different considerations led *Wilser* (1928), and *Muratov* and *Neprochnov* (1967) to conclude that the central part of the Black Sea was flooded at the beginning of Late Cretaceous time. With the Pyrenean phase, the Black Sea became the foredeep of the Outer Pontides (p. 75), and started to sink more rapidly. Sedimentation in the Black Sea proper, therefore, probably began with Upper Cretaceous strata and consists mainly of Oligocene, Neogene, and Quaternary deposits. A 50 km-wide belt along the Anatolian coast shows stronger magnetic disturbances (Anom., 1965, Plate 5; *Ross* et al., 1974: 27), it may belong to the Outer Pontides. The central part of the sea is free of earthquakes, and the young sediments are flat. It is not known when euxinic conditions started in the Black Sea. They may have occurred at an early stage, if the sedimentary deposits of oxidic manganese ore in the North Anatolian Upper Cretaceous (*Chazan,* 1947), and the Thracian Oligocene (*Uzkut,* 1971) can be regarded as a sign of a decrease in oxygen in the basin itself.

The thickness of the clastic deposits in the geosynclines to the north and south of the Black Sea allows to estimate the denudation of the Pontian land during the Paleozoic and earlier Mesozoic periods to at least 10 km. The thinning of the granitic layer beneath the Black Sea could, thus, find its explanation. The reduction in thickness of the basaltic layer, however, cannot have happened in this way. *Khain* (1964) attributed it to crustal expansion. This seems reasonable, especially since the emplacement of the ultramafitite massifs in Anatolia can best be explained by expansion (p. 122). The present area of suboceanic crust in the Black Sea is comparable to the Middle Anatolian ophiolite zone in its dimensions and its east-west extension. Both formed at approximately the same time during the Late Mesozoic.

The Northern and Central Aegean Sea (Fig. 55)

The crust in the Aegean Sea is approximately 30 km thick, and thus thinner than in adjoining Anatolia (*Makris,* 1973). According to the mostly positive free-air gravity values and the fluctuating pattern of the magnetic isanomals, the basement is composed of metamorphites and magmatites, and thinly covered by sediments. At the bottom of the sea are drowned valleys and extensive remains of the Thracian peneplain (p. 70). The Wallachian faults of the West Anatolian coastal area continue on the sea floor. The fault line which separates the Kaz Dağ from the Gulf of Edremit can be pursued as far as the Island of Skiros. The graben of the Kücük Menderes opens seaward to the deep bay between Karaburun-Chios and Samos-Ikaria. The Aegean Sea thus represents a former land that was flooded only recently, and that shows the same block structure as in West Anatolia. However, tectonic movements in the central part of the Aegean Sea ceased, for it is almost free of earthquakes.

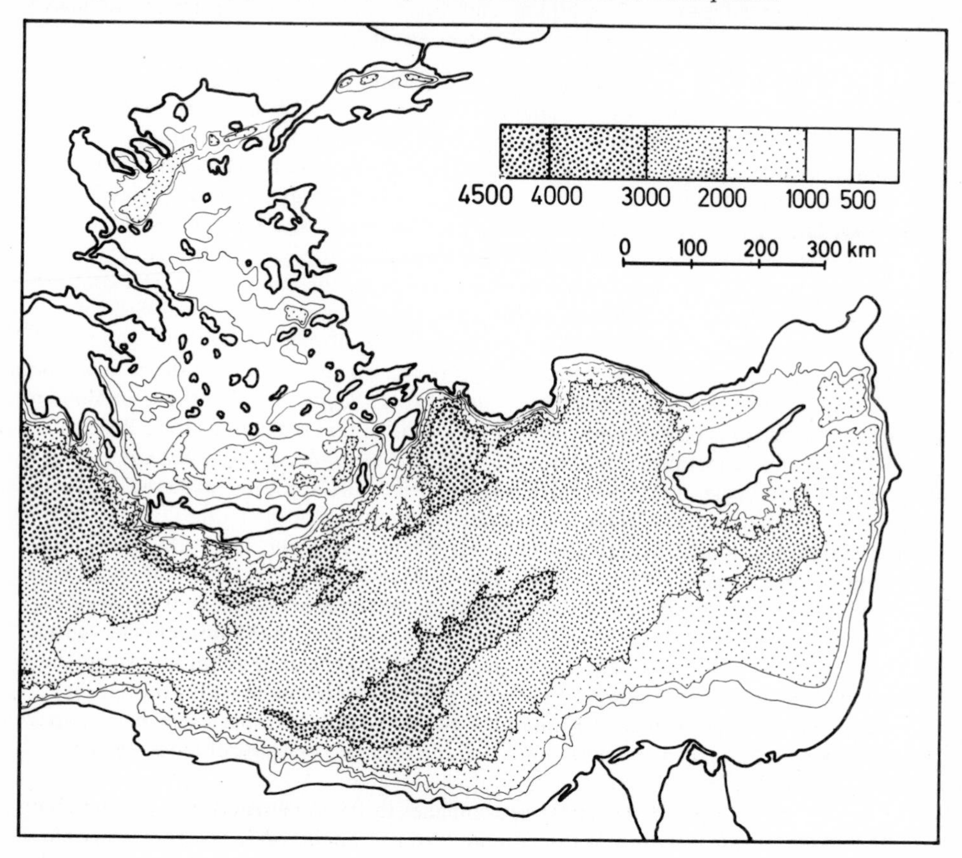

Fig. 55 Depth of the Levantine Sea and of the Aegean Sea (after *Goncarov* et al., 1966)

In the northern part of the Aegean Sea a chain of troughs with depths of up to 1600 m extends from the Gulf of Saros to close to the Greek coast. The troughs accompany the western continuation of the North Anatolian seismic line (*Jacobshagen,* 1972: 451; *McKenzie,* 1972: 144). Based on gravity, the trough between Samothraki and Limnos is a tectonic graben. The trough north of Skopelos, however, is characterized by a gravity-plus, magnetic disturbances, and an increased heatflow (*Needham* et al., 1972). With regard to the relations between seismic lines and ophiolite zones (p. 100), it is possible that off Skopelos a tectonically and magmatically active ophiolite trough exists.

The Southern Aegean Sea and the Levantine Sea (Fig. 55–57)

In the sea south of the Cyclades, negative free-air gravity values are prevalent. The minor magnetic disturbances indicate thick sediments. The crust of 25–27 km is thinner than in the Aegean Sea (*Lort* et al., 1974: 365).

The islands and depths are arranged in several parallel arcs, convex to the south. From north to south they consist of:

1. A chain of young volcanoes of which the Island of Thira (Santorin) has been active until recently (*Ninkovich* and *Hays,* 1972; *Günther* and *Pichler,* 1973). On the Anatolian mainland, this chain may continue as far as Kula east of Izmir (*Philippson,* 1913).
2. The trough of Crete which is up to 2500 m deep.
3. The ridge of Crete which is outlined by the Isles of Crete, Karpathos, and Rhodes. It connects the mountain ranges of Greece and the western Taurus.
4. The 2000 to 5100 m-deep Hellenic trough. It coincides with a belt of -100 to -150 mgal gravity. Several seamounts rise on its southern rim. They are, like Crete, gravimetrically positive.
5. The Mediterranean ridge. It extends from southern Italy to Cyprus as a 150 km-wide, 700 m-high welt and is accompanied by a free-air gravity minimum of approximately 150 mgal. In the Levantine Sea, the ridge is composed of thick beds of young sediments (*Lort* et al., 1974: 365). In the Ionian Sea, however, it has a core of crystalline rocks, according to *Hinz* (1974: 49).
6. The Herodotus Basin, a deep-sea plain which is situated off the coast of Egypt.

From Miocene to Pannonian time, marine sediments were deposited in the Levantine Sea; however, in a progressively restricted environment. The Pannonian evaporation period is mainly represented by brackish deposits. Evaporites have not been found at the bottom of the Levantine Sea yet, but are probably present (*Hsü* et al., 1973). In the late Pliocene, normal marine conditions became reestablished. Since that time, mostly calcareous, fine-grained turbidites, intercalated with sapropelite and volcanic ash, are deposited. The clastic material is mostly provided by the Nile. The carbonate fraction consists of planktonic organisms (*Stanley,* 1972; *Keller* and *Ninkovich,* 1972). The young sediments in the trough of Crete, in the Hellenic trough, and on the northern slope of the Mediterranean ridge are disturbed, in part by slumping, and in part by Holocene tectonic movements (*Wong* et al., 1971; *Stanley,* 1973).

The seismic phenomena fit into the zonal structure. Shallow-focus earthquakes are particularly frequent in the Hellenic trough. The deeper foci, however, are gradually shifted north (*Papazachos,* 1973). This arrangement is very distinct in the southern Aegean Sea, but can be traced, less distinctly, along the southern coast of Anatolia to Hatay. Focal mechanisms point to compression perpendicular to the direction of the arc (*McKenzie,* 1972: 144).

Geological and geophysical investigations show that a tectonic line dissects the northern Levantine Sea from west to east. In its western part, it represents a Benioff zone following the Hellenic trough and dipping toward the Aegean Sea at an angle of 30° to the north. In the east, it forms the southern boundary of the ophiolite zones of Southwest Anatolia, Cyprus, and Hatay; it can also be traced magnetically (*Kogan* et al., 1969; *Lort* and *Gray,* 1974).

Rabinowitz and *Ryan* (1970: 600) assume that the thickness of sediments in the Mediterranean ridge is 5 to 7 km greater than in the rest of the Levantine Sea. They explain this by repeated imbrication during the underthrusting of the Levantine Sea plate under the arc of Crete. Geologic considerations, however, suggest an active nappe thrust towards the south in the arc of Crete. This is supported by the slump and redeposition phenomena observed

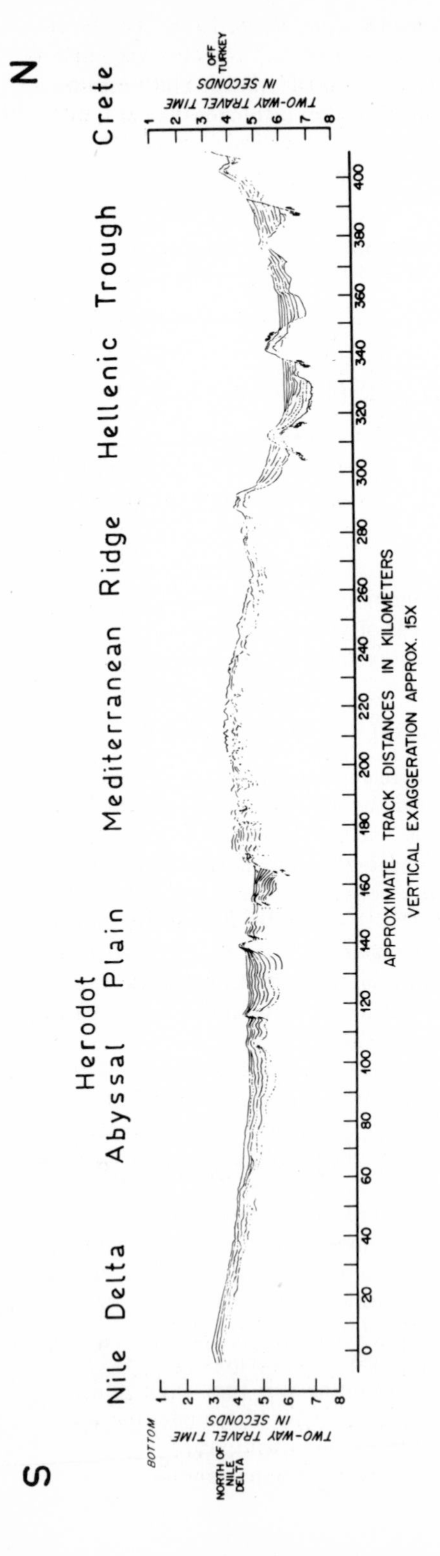

Fig. 56 Seismic profile across the Levantine Sea from Southwest Anatolia to the Nile Delta. Vertical exaggeration about 15 x (after *Wong* et al., 1971)

in JOIDES cores in the southern part of the Hellenic trough (*Hsü* and *Ryan*, 1973). The Mediterranean ridge can be interpreted as the fill of a late Neogene-Quaternary foredeep accompanying the Crete ridge. It did not begin to rise until most recently, probably in connection with the nappe movements (*Ryan* et al., 1973b: 371). In the northwestern part of the trough, different conditions seem to prevail; here, *Hinz* (1974: 58) found only young vertical displacements.

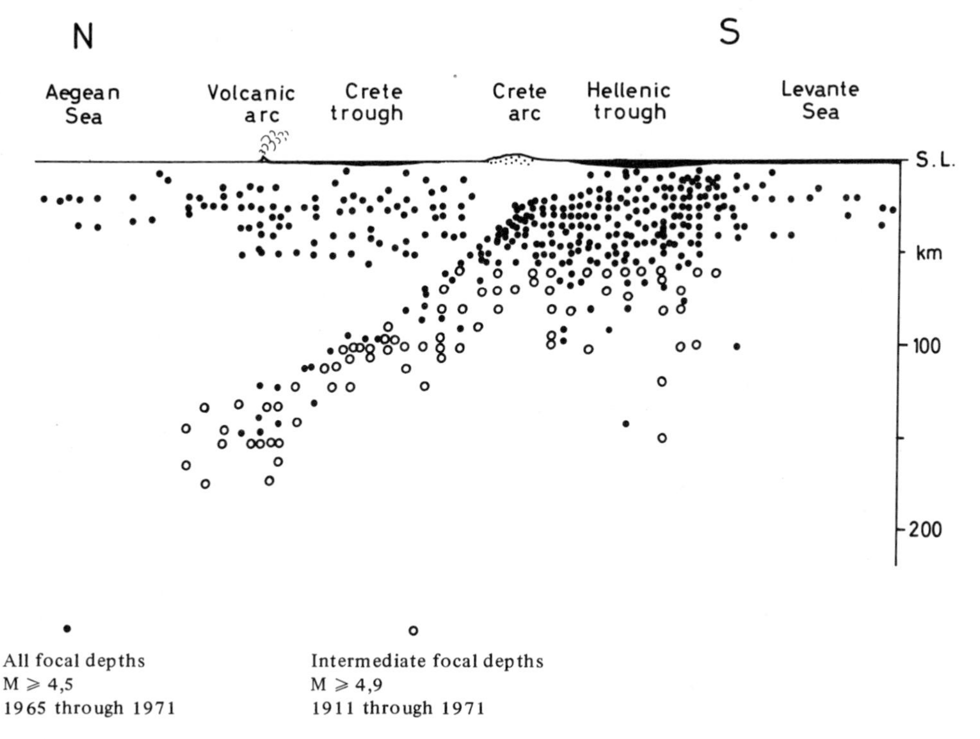

All focal depths
$M \geqslant 4{,}5$
1965 through 1971

Intermediate focal depths
$M \geqslant 4{,}9$
1911 through 1971

Fig. 57 Section of the Benioff zone in the subsurface of the southern Aegean Sea and the northern Levantine Sea near Crete (after *Papazachos*, 1973)

Chapter 21 The Geotectonic Classification of Turkey

The various ideas on the geotectonics of Turkey have been discussed by *Leuchs* (1943). More recent contributions have been provided by *Arni* (1939a), *Egeran* (1945), *Blumenthal* (1946), *Ketin* (1966a), and *Ilhan* (1971a).

The Geotectonic Classification in the Paleozoic (Fig. 14)

The Eastern Mediterranean and Near East regions fall into three major geotectonic units which remained through the entire Paleozoic. To the north, the Pontian land represented a craton which had been rising for a long time. It was separated from the East European platform by the Paleozoic geosyncline Dobruja-Crimea-Greater Caucasus. The Pontian land occupied the inner part of the Black Sea, but it overlapped into the present mainland by the Moesian platform and the Transcaucasian Massif. The Afro-Arabian shield was to the south. In Cambrian-Ordovician time, its northern edge was the Sinai-Jordania line. During the course of the Paleozoic, it shifted to the north, and during the Carboniferous-Permian period, it crossed the Levantine Sea and Syria (*Wolfart,* 1967b: 179).

The area between these two cratons was occupied by the Tethyan geosyncline. Even in the early Paleozoic, its floor was composed of units with differing mobility. On the one hand, there were highly mobile zones with stratigraphically complete profiles from the Cambrian to the Devonian, such as the Sultan Dağ, the Antitaurus, and Southeast Anatolia. On the other hand, there were semi-consolidated areas with transgressive Devonian strata, such as near Bolu, Kayseri and Anamur. They represent the first outline of the North Anatolian welt, the Kirşehir, and the Alanya Massif. The Variscan mountain building added to this inhomogeneity, even though the cratonization following the folding was remarkably less intense than in Central Europe. Only a few blocks, such as Kocaeli, the apparent Variscan central zone from the Biga Peninsula to Ankara, and the Sultan Dağ were completely consolidated. Other areas, such as the eastern Taurus and the Antitaurus, were subjected to a partial consolidation.

Variscan Mountain Connections (Fig. 14, 58)

So far, the Variscan mountains in Turkey and in Central Europe can be linked only cautiously, as the Variscan structures in the intermediate area are not sufficiently known. No controversy exists concerning the northern boundary of the tectogene toward the East European platform (*Belov,* 1967; *Muratov,* 1969; *Von Gaertner,* 1969). But the southern boundary in Alpidic Southeast Europe is uncertain. Here, a clue on the extent of the Variscan folded area is provided by the distribution and facies of the Late Paleozoic rocks. Where Carboniferous and Permian strata are developed as marine limestones, they are usually conformable with the older Paleozoic as well as with the Triassic sediments, such as in large parts of the Dinarides. The denudated Variscan mountains, however, are mostly unconformably covered by Permian rocks of Rotliegende facies and by Triassic strata with Germanic affinities such as in Bulgaria (*H. Flügel,* 1975). Between the Black Sea and the Caspian Sea, the southern boundary of the Variscan mountains can be deduced by the observation that in the northern part of the Greater Caucasus, the older Paleozoic rocks are unconformable and on the southern slope are conformable to the upper Paleozoic and Mesozoic strata (*Somin* and *Belov,* 1967).

The Turkish Variscides fit this frame in the following way. The supposed central zone of the tectogene consists of crystalline rocks of the North Anatolian welt and the Serbo-Macedonian Massif (*Kockel* et al., 1971). It is accompanied by external, sedimentary zones on either side. The northern, north-verging zone runs from southern Bulgaria to the Anatolian Black Sea coast. Sections across the Variscan folds in Southern Bulgaria (*Tenchov,* 1971) and in Northwest Anatolia are very similar. It is probable that northeast of Inebolu the northern Variscan external zone continues on the floor of the Black Sea. The Jurassic beds on the Crimea contain granitic pebbles with an age of 280 to 210 m. y. which stem from the Pontian land (*Milanovsky,* 1967: 1241). The southern external zone, which in the Sultan Dağ is verging to the southwest, borders the Inner Anatolian Massif in the south. The zone may continue via the Bitlis Mountains into the central part of Iran. Here, the observations on late Variscan unconformities are still controversional (*Stöcklin,* 1968a: 1239; *Thiele,* 1973).

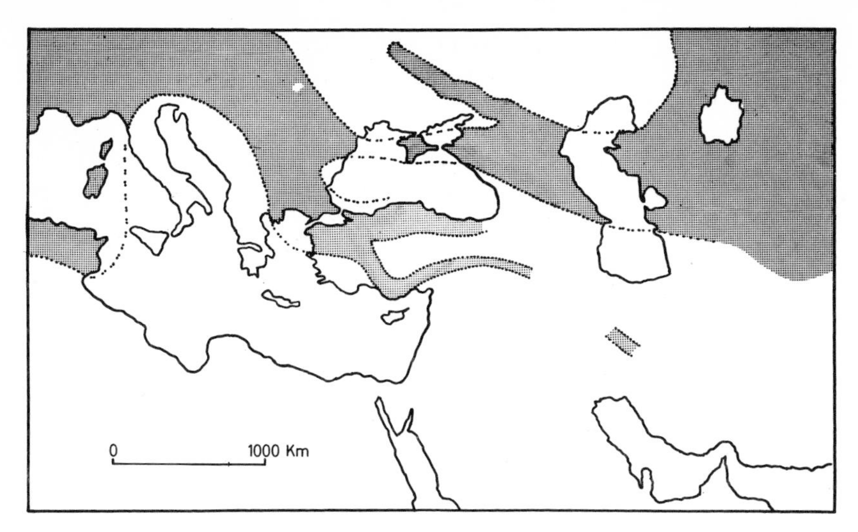

Fig. 58 Area of Variscan mountain building in the Eastern Mediterranean and Near East regions

Alpidic Geotectonic Units (Fig. 59, 60)

The present geotectonic map shows a symmetric structure. In the north and south of Turkey, inland seas, such as the Black Sea and the Levantine Sea, are covering the margins of the old cratons, the East European platform, and the Afro-Arabian shield. Young Alpidic mountain ranges accompany the northern and southern coast lines. In Thracia and Anatolia, they comprise a „zwischengebirge" which consists of units of different age. From north to south the following structural entities can be recognized:

1. During the middle Cretaceous epoch, the Pontian land turned into an epicontinental sea, and starting with the Oligocene, into a foredeep for both the Pontides and the Crimea-Caucasus tectogene. The remobilization of this former cratonic block is still going on.

2. The Pontides were formed in two phases. The older mountains, the Kimmerian Inner Pontides, are not very distinct, as their greater part became incorporated into the Outer Pontides.

3. The „zwischengebirge" poses many open questions. In the order of their consolidation ages, the following parts are distinguishable:

   a) Contrary to present theories, pre-Alpidic consolidated areas are rather widespread. The basement of the Thracian Basin, the Sea of Marmara, as well as the Inner Anatolian Massif may have existed as cratons since Cryptozoic time. At the end of the Paleozoic, parts of the Variscan mountains became consolidated (p. 114).

   b) The ophiolite zones and the areas with Mesozoic regional metamorphism are regarded as early to middle Alpidic consolidation areas, because they started to appear as uplift areas in the later Mesozoic.

4. The mountains of the western Taurus and Southeast Anatolia underwent two tectogenic phases, but on the whole, they are somewhat younger than the Pontides.

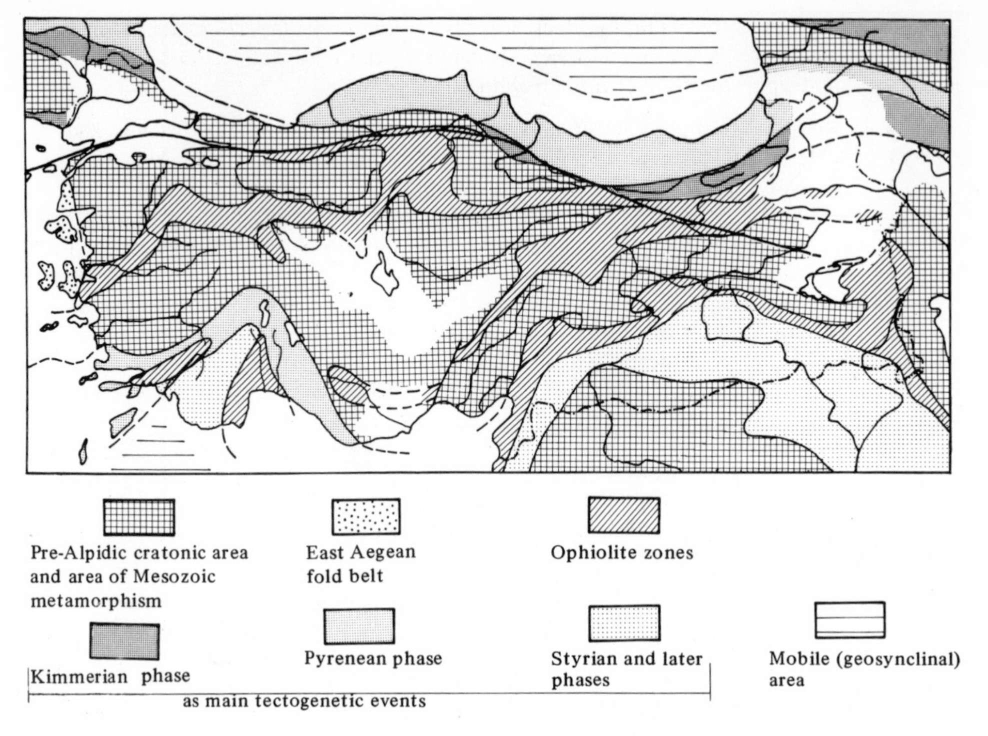

Fig. 59 Geotectonic map of Turkey

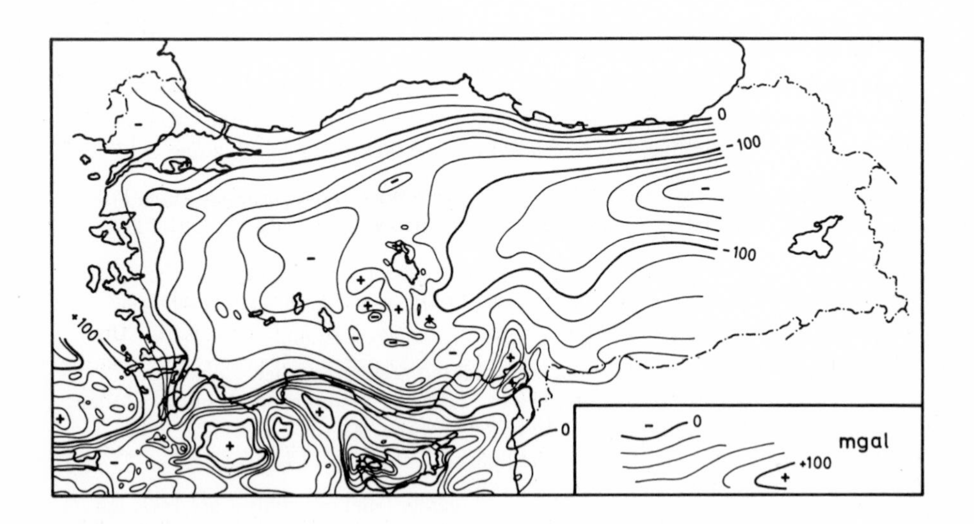

Fig. 60 Bouguer gravity anomalies in Turkey (after *Özelçi,* 1973)

5. In the Levantine Sea as well as in the Black Sea, thick series of young sediments accumulated on an old craton. The basin and its extension along the Turkish-Syrian border region represent a foredeep in the state of initial folding.

In this way, almost all mobile zones of Turkey were turned into Alpidic tectogenes, and have been consolidated. Only few potentially foldable areas remained. The Thracian and the Tuz Gölü basins are situated on two of the oldest cratons. Their fill, which measures more than 3000 m, is still horizontal. Also the western part of the basin of Adana-Iskenderun is undisturbed. While the mainland consolidates progressively, new mobile areas are generated in the Black Sea and the Levantine Sea proper.

The gravity map of Turkey (*Özelçi*, 1973) shows clearly to what extent Anatolia has become a cratonic block and how its crustal structure differs from that of its surroundings. Middle Anatolia shows consistently negative values of 50 to 150 mgal. This means crustal thicknesses well above the normal ones. The Bouguer zero isanomal closely follows the outline of the mainland. At the Black Sea and Levantine Sea coast, gravity increases rapidly, and at the Aegean coast more slowly to positive values. Due to the wide spacing of the gravimetric measurements, only few distinctive features can be seen on the map. The young subsidence areas of the Tuz Gölü area and of the Bay of Adana are characterized by relative maxima. The narrow gravity minimum from Afyon across the Sultan Dağ to the Antitaurus coincides with the southern sedimentary zone of the Variscan mountains (Fig. 14). It could be explained by a belt of Paleozoic rocks thickened by folding. The minimum zone of Crete-Rhodes-Southwest Anatolia represents an analogous structure of Alpidic age, connecting the Hellenides with the Taurides. Farther to the east, from Antalya via Cyprus to Hatay, the Taurus tectogene is accompanied by positive anomalies of up to 200 mgal. They can be attributed to the ultramafics prevalent in that area.

The geotectonic development is also expressed in the gradual replacement of the flysch facies by the molasse facies. In the Pontian geosyncline, the flysch is present until the Eocene, in the Taurian geosyncline until the middle Miocene, and at Adana and on Cyprus until the late Miocene. On the floor of the Levantine Sea, flysch-like sediments are being deposited even today. The molasse first appeared in Middle Anatolia at the turn of the Cretaceous (p. 57). Then it migrated toward the north and south, and replaced the flysch. In Thracia it appeared in the Eocene, and at the northern and southern coasts of Anatolia in the Miocene; at Iskenderun its deposition continued till the Quaternary.

A comparison of the geotectonic classification proposed above shows agreements and differences with the schemes of other authors:

1. The term „Pontides" (*Arni*, 1939a) is used, as far as Northeast Anatolia is concerned, in a similar sense as *Ilhan* (1971a) and *Ketin* (1966a). There are differences in Northwest Anatolia. According to *Ketin*, the Pontides widen toward the west and comprise the Ulu Dag, Kaz Dağ, and the Karaburun Peninsula. But there is evidence that neither the Kocaeli nor the Biga Peninsula show a complete geosynclinal sequence or an alpine-type structure. Observations support the view that the Pontides leave the mainland west of Ereğli, as suggested by *Arni*. The shoals on the floor of the Black Sea off the Bosporus may mark their continuation (*Ross* et al., 1974: 19).

2. The Anatolides (*Arni*, 1939a) coincide almost with the above-mentioned zwischengebirge.

3. The term „Taurides" (*Arni*, 1939a) should be limited to the western Taurus, after *Blumenthal* and *Brunn* et al. have shown the geosynclinal and tectogenic individuality of these mountains. This poses the question of how the eastern Taurus, the Antitaurus, and the environs of the Van Gölü should be interpreted geotectonically. Maybe these areas were subjected to a partial consolidation by the Variscan folding, so that they were less touched by the Alpidic tectogenesis. However, new investigations in this little-known region are needed.

4. In Southeast Anatolia, *Arni's* classification (1939a) into an inner thrust zone and an external fold zone, which he combined under the term „Iranides", is still valid today (*Ten Dam,* 1965).

Alpidic Mountain Connections (Fig. 61)

The young tectogenes of Turkey continue into the neighboring countries. Starting from the north, the following relations exist:

1. The Outer Pontides are linked with the Bulgarian Srednogorie (*Wilser,* 1928: 211) to the west, and with the zone of Adjar-Trialet in Armenia (*Khain* and *Milanovsky,* 1960/63: 677; *Gabrielyan,* 1964, 1968) to the east. All three areas have a volcanic flysch facies in the Late Cretaceous and Early Tertiary, and were folded during the Pyrenean phase. Sulfidic copper mineralization is common to all three regions.

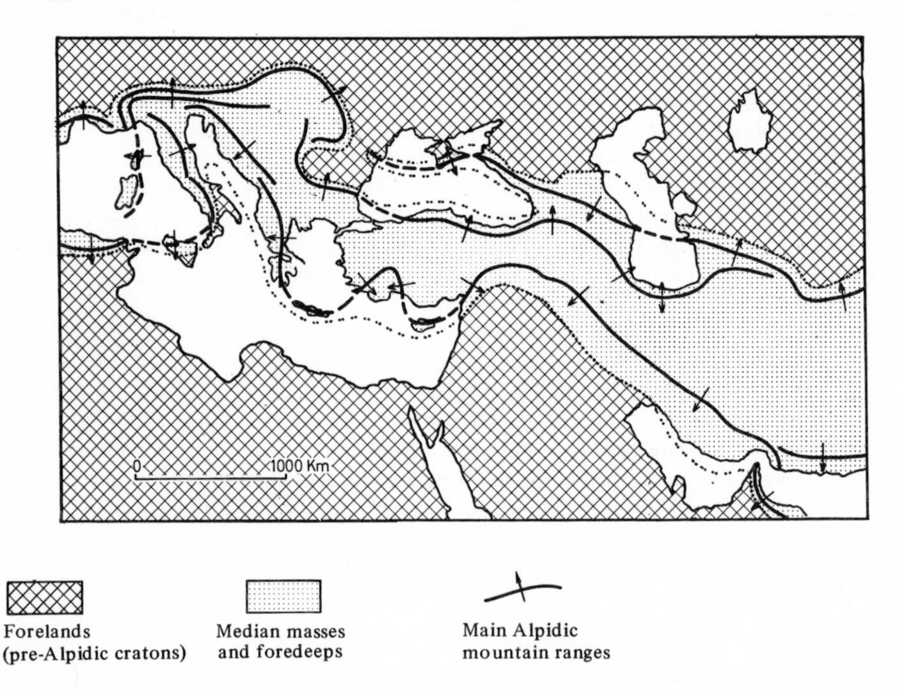

Fig. 61 Geotectonic map of the Eastern Mediterranean and Near East regions (partly after *Khain,* 1968, and others)

2. Strong Kimmerian phases are a special characteristic of the environments of the Black Sea; they are missing in the remainder of the Mediterranean region. The connection between the Inner Pontides of Northeast Anatolia with the Istranca and the south Bulgarian Sakar-Strandca zone has already been substantiated (p. 52). In the east, close relations exist to the zone of Somkhet-Kafan (*Khain* and *Milanovsky,* 1960/63: 677; *Gabrielyan,* 1964, 1968). As in Anatolia, this zone is mostly covered by Malm and Cretaceous sediments and has become part of younger mountain ranges. The Dobruja-Crimea-Greater Caucasus tectogene north of the Pontian land is the counterpart to the Inner Pontides. Thrusting and plutonism took place during the same time, except that they were much more intensive in the north of the Black Sea. In the Greater Caucasus, Upper Dogger sediments are overlying older

Jurassic rocks in angular unconformity. The age of the granites and diorites lies between 175 and 163 m. y. (*Rubinstein,* 1970: 286). From both sides, the vergence is toward the Black Sea.

3. Most authors link the Middle Anatolian ophiolite zone with the Vardar zone via Izmir (*Brunn,* 1960: 472). On the Aegean islands in between, however, no traces of ophiolitic rocks are known. On the other hand, there are definite ophiolite occurrences in the Tekirdağ (*Kopp* et al., 1969: 48) and questionable ones in the northern part of the Aegean Sea (p. 110). All these indications, together with the results by *Kockel* et al. (1971: 549), show that the Middle Anatolian zone probably joins with the Vardar zone via Bolu, Mudurnu, and the Tekirdağ. In Armenia, the ophiolite sequence of the Middle Anatolian zone reappears in the Sevan-Akera zone east of Lake Sevan. According to *Knipper* (1971), it is of Early Cretaceous-Jurassic age, or possibly older, and was folded during the Austrian phase.

4. The north-trending structures of the West Anatolian coastal area (p. 92) are also present at Chios (*Besenecker* et al., 1968: 140) and Lesbos (*Hecht,* 1972). *Philippson* (1914) attributed all of them to an East Aegean fold belt. But today, this idea is abandoned (*Richter,* 1966: 77; *Jacobshagen,* 1972: 452).

5. In the south of Anatolia, the Taurides of the western Taurus continue, on the one hand, into the Hellenides via the arc of Crete (*Bernoulli* et al., 1974), and on the other hand, via Cyprus and Southeast Anatolia into the Zagros thrust zone (*Stöcklin,* 1968a: 1246). These relations have been known for quite a while (*Arni,* 1939a: Plate 2). The connection Taurus-Cyprus-Hatay is further supported by stratigraphic and facial analogies as well as by geophysical data in the northern part of the Levantine Sea (*Vogt* and *Higgs,* 1969: 441; *Woodside* and *Bowin,* 1970: 1111; *Lapierre* and *Parrot,* 1972; *Lort* and *Gray,* 1974).

In conclusion, the eastern part of the Mediterranean region and the Near East are crossed by three young tectogenes present as mountain chains. The northern chain, Dobruja-Crimea-Greater Caucasus-Balkhan-Kopet Dağ, extends outside Turkey. The other two chains accompany the northern and southern coast of Anatolia. They comprise a zwischengebirge which consists of a mosaic of blocks that had turned into cratons in middle Alpidic, early Alpidic, Variscan, and even earlier times.

Chapter 22 Some Geotectonic Problems of Turkey

The Ophiolite Zones and Plate Tectonics (Fig. 62, 63)

Plate tectonics is based on a Tethyan sea which in the Near East was approximately 2000 km wide during the Paleozoic. From the Liassic, the Atlantic Ocean opened and received its present outline, predominantly during the Cretaceous. At the same rate, Eurasia and Afro-Arabia came closer and encroached on the Tethys. The two continental plates depressed the intermediate oceanic plate until they touched. In Turkey, the suture between the two continental plates should be visible in the field as a tectonic lineament running east-west throughout Anatolia. Different alignments are conceivable.

According to *Dewey* et al. (1973), the Anatolian plate was situated at the southern margin of the Tethys during the late Paleozoic. In that case, the suture between the Eurasian and Afro-Arabian plate should follow the North Anatolian coastline. *Smith* (1971: 2048) thus considers the Pontides a sedimentary bulge that was squeezed above the subduction zone. This assumption severs the Variscan tectogene (Fig. 58) into two discontinuous units on either side of the Tethys (Fig. 62). It is also hard to reconcile with paleobiogeographic data. The Paleozoic marine fauna as well as the Late Carboniferous land flora of Anatolia show close ties to Central Europe (p. 27, 32). During the Jurassic period, northern Anatolia and the area of the Caucasus formed one province (p. 52). The same holds true for the Late Cretaceous epoch (p. 61). In the Pontian geosyncline, the Upper Jurassic to Eocene strate are nearly conformable; they were not folded intensively until the Pyrenean phase. The suggested suture, thus, was tectonically almost inactive during the late Mesozoic – the major epoch of the subduction – and was not activated until the early Tertiary. On the other hand, the available paleomagnetic data favor a subduction along the Pontides (*Zijderveld* and *Van Der Voo,* 1973). The Anatolian pole positions coincide with the Afro-Arabian ones but can only be reconciled with the Eurasian pole positions by a counterclockwise rotation of Anatolia by 50 to 80° (Fig. 63). The investigated rock samples range from the late Carboniferous to the Eocene, and originate from the central and northern part of Anatolia. Accordingly, all of Anatolia should have belonged to the Afro-Arabian plate since the late Paleozoic.

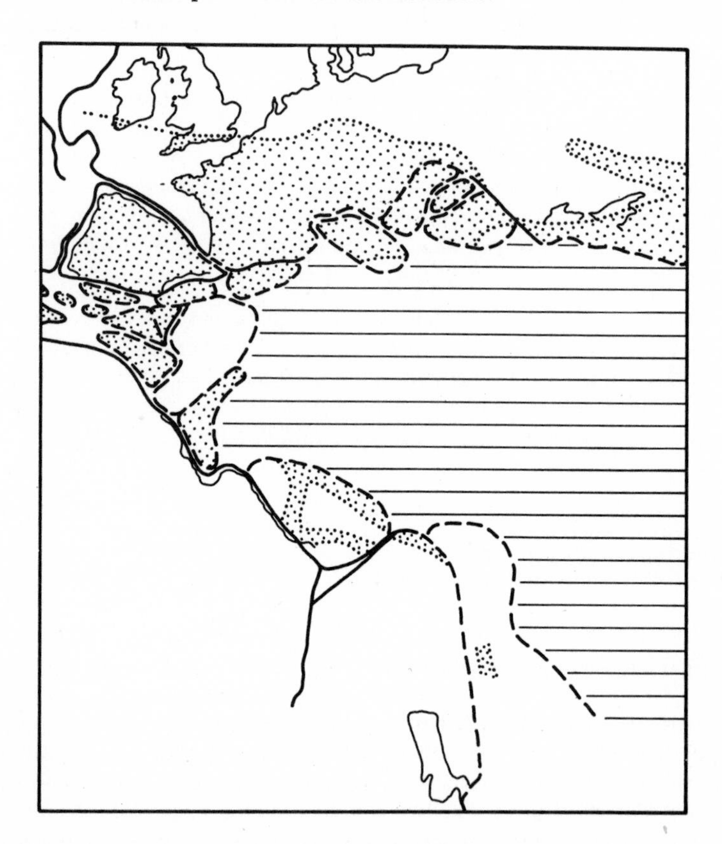

Fig. 62 The Mediterranean and Near East regions at the end of the Paleozoic era according to the theory of plate tectonics (*Dewey* et al., 1973). Variscan fold belt stippled (compare Fig. 58), Tethys ocean ruled

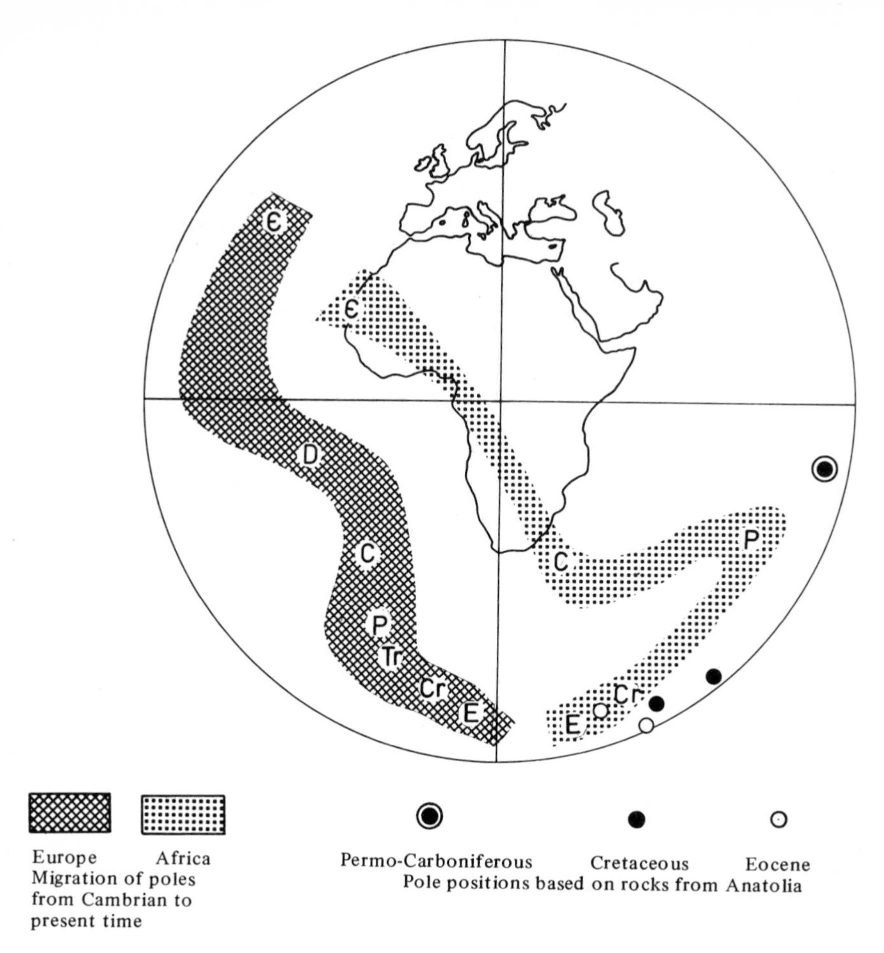

Fig. 63 Migration of paleomagnetic poles in Africa and Anatolia in relation to stable Europe (after *Zijderveld* and *Van Der Voo,* 1973)

Another possibility is to consider the ophiolite zones as plate sutures. The zones contain rocks of oceanic provenance, such as radiolarite, basalt and ultramafitite which are completely missing in the Pontides. But again, there are discrepancies. In Anatolia, two east-west trending ophiolite zones exist. The Middle Anatolian zone can be traced from the Aegean coast to Armenia. A southern zone runs from the western Taurus via Cyprus to Southeast Anatolia and Iran. *Horstink* (1971) assumes two subduction zones and considers southern Anatolia a microplate between the two large plates. But according to *Enay* (1975), the two ophiolite zones differ in biogeographic function. The Middle Anatolian zone separates two faunal realms in the Jurassic (p. 52), and should represent the suture of the wide Tethys ocean. The southern zone, in contrast, does not constitute a faunal limit.

The spatial and temporal distribution of the Turkish ophiolite zones has been described on p. 87. Their structure and rocks were investigated in detail on p. 84. Now the question remains, whether they can be regarded as subduction zones according to their nature and geotectonic position. Anatolia, in particular its central part, is especially well situated for a starting point. Here, the ophiolite zones have not been subjected to pronounced tectonic deformation in post-Mesozoic time, and their original structure is well preserved (*Hsü,* 1971; *Kazmin,* 1971; *Knipper,* 1971; *Abbate* et al., 1972; *Dimitrijević,* 1973; *Moores,* 1973; *Juteau* et al., 1974). The following points are raised:

1. The lithology of the Anatolian ophiolite sequences is indeed similar to that of the oceanic crust (*Gass* and *Smewing,* 1973). The upper, mostly sedimentary part can be compared with crustal layer 1, the deeper part, which contains more volcanic rocks, with layer 2, and the ultramafic basement with crustal layer 3. The thicknesses, however, differ strongly. The equivalents of layers 1 and 2 in Anatolia attain a maximum thickness of 1 km, i. e. reach only a fraction of their thickness present on the ocean floor. In parts, such as between Erzincan and Tarsus (p. 49), neritic beds overly the ultramafitite directly; however, this may be due to subsequent erosion.

2. A detailed study reveals also material differences between the ophiolite zones and the ocean floor. According to the data available from Anatolia (Fig. 30), the ophiolitic lavas are mostly alkaline-basaltic, more rarely tholeiitic (*Bailey* and *Blake,* 1974). In addition to radiolarite and pelagic limestone, the sedimentary rocks are composed mostly of turbidite and slump deposits (p. 85). Their facies indicates deposition in abyssal troughs near to the coast or in a deep sea interspersed with islands, but hardly in an open Tethys of oceanic dimensions.

3. In accordance with the gradual encroachment of the Tethys by the progreding subduction, Permian, Triassic, Jurassic, and Cretaceous strata should be present in the stratigraphic column of the ophiolite sequences in decreasing fraction. Even the present limited observations show a different picture. A Permian ophiolite facies has so far not been found; however, this facies was especially widespread during the Cretaceous. Some ophiolite sequences in Anatolia and the neighboring countries did not last longer than the Malm (p. 87; *Brunn* et al., 1972b). This would suppose considerable subductions at a very early time.

4. Facies and areal distribution of the Mesozoic sedimentary rocks of Turkey give no indication that the country consists of two or three plates that once were far apart. The ophiolite zone do not coincide witz lithofacial boundaries. On the contrary, their rims show a good fit.

5. Sections across subduction zones generally show an asymmetrical structure with inclined dislocations. In contrast, the Middle Anatolian ultramafitite massifs have steep contacts. Belts of volcanoes or mineral provinces that are in spatial or temporal connection with the supposed subduction zones, have not been found so far. Thrust planes are only present in the ophiolite zones of South Anatolia where the ophiolitic rocks have been included in alpine-type tectogenes long after their eugeosynclinal development had come to an end.

6. *Ernst* (1973) has explained the formation of lawsonite-glaucophane rocks by metamorphism in the descending plate. Undoubtedly, a resemblance exists between the Californian Franciscan (*Maxwell,* 1974) and the Turkish ophiolite zones. In Anatolia, however, unmetamorphosed ophiolitic rocks are in close association with metamorphosed ophiolitic rocks (p. 86); a fact, which is hard to explain by subduction.

The Ophiolite Zones as Rift Structures (Fig. 64–66)

An alternate hypothesis for the ophiolite zones is crustal expansion, and the openings are filled with rocks from the lower lithosphere. This hypothesis is favoured for the following reasons:

1. The intrusion of rock masses with dimensions of tens of kilometers is only possible by dilation of the crust, at least at the beginning stage of the rise of the ultramafitite. The veins of gabbro, dolerite, and basalt which steeply dissect the ultramafitite massifs in the form of parallel swarms (*W. J. Schmidt,* 1954; *Vuagnat* and *Çoğulu,* 1968), also require a tangential expansion.

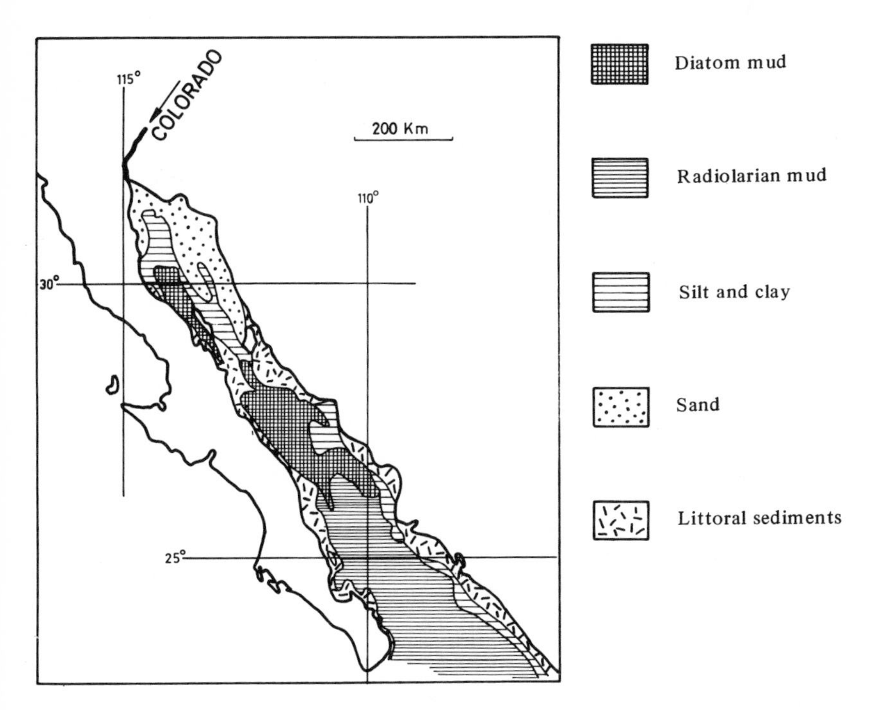

Fig. 64 Recent sediments of the Gulf of California (after *Van Andel,* 1964)

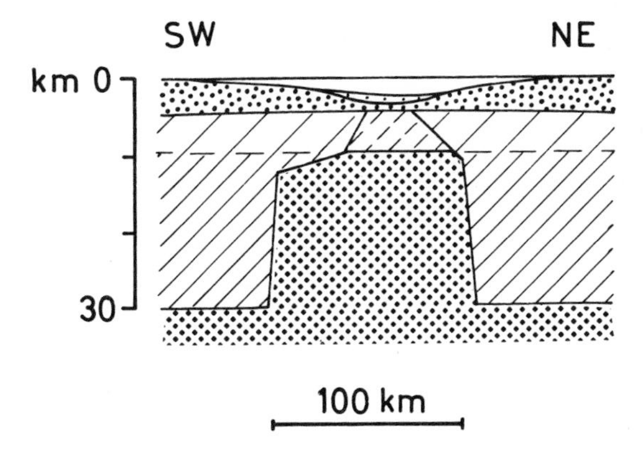

Fig. 65 Section across the Gulf of California according to geophysical measurements (after *Harrison* and *Mathur,* 1964. For explanation see Fig. 54)

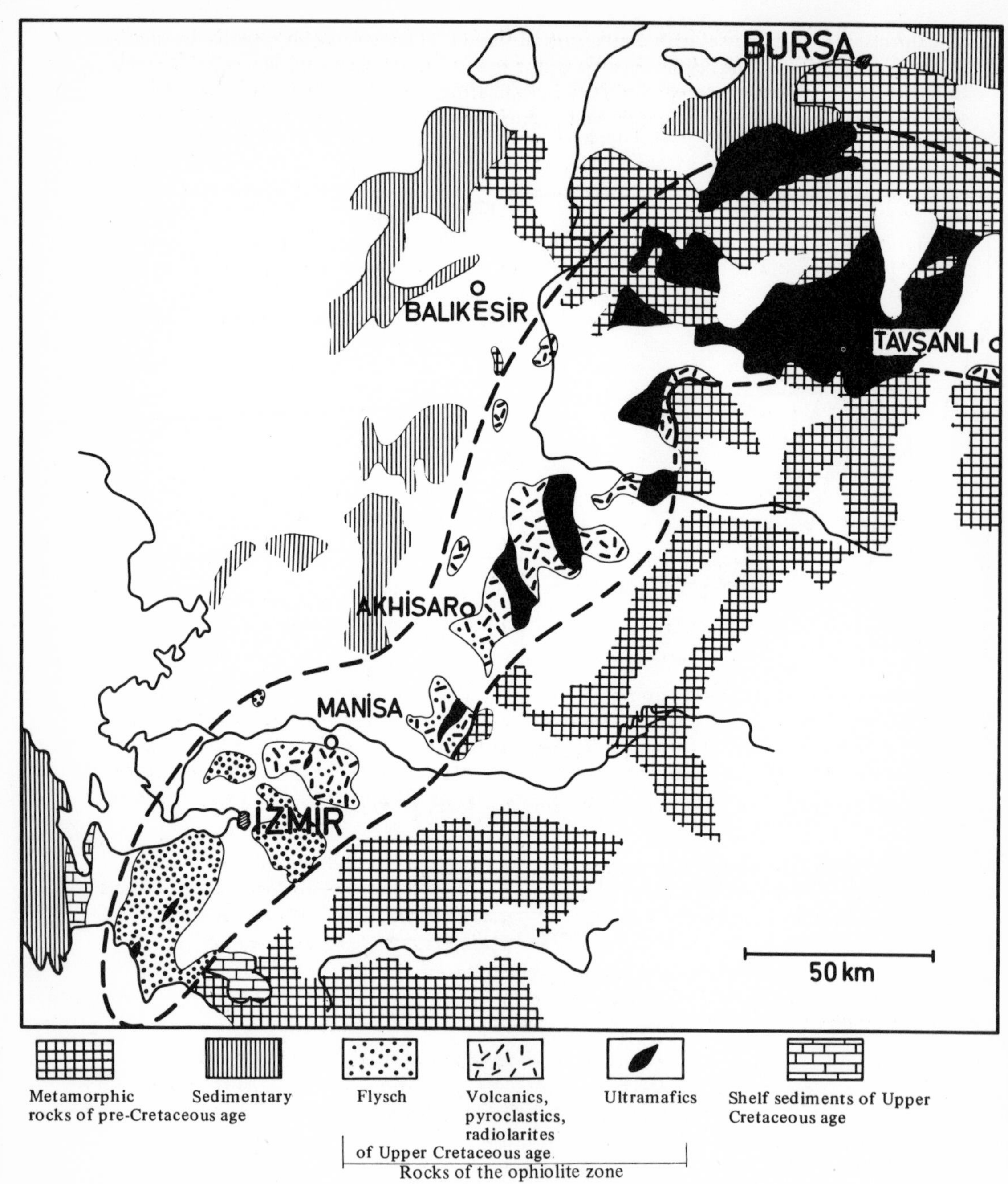

Fig. 66 Western part of the Middle Anatolian ophiolite zone between Izmir and Tavsanlı

2. A second argument is the similarity of the ophiolite zones with structures generally related to expansion. As an example, the comparison of the Gulf of California with the western part of the Middle Anatolian ophiolite zone is cited. The formation of the Gulf can be attributed to the separation of Baja California from the North American continent in conjunction with a horizontal displacement along the San Andreas fault (*Rusnak* et al., 1964: 74; *Moore*, 1973). Both movements did not start until fairly recent and are still continuing

(*Molnar,* 1973). In the subsurface of the Gulf a rift zone en echelon opened up. In the south, it widens to a width of 100 km. In the north, it narrows down, and at the end of the Gulf, it continues as the San Andreas line. According to geophysical data, the rocks that fill the fissure correspond in part to those of the oceanic crust, and in part to those of the mantle (*Harrison* and *Mathur,* 1964; *Phillips,* 1964). Topographically, the rift zone appears as a chain of ocean basins whose depths increase with the widening of the fracture to the south. In the north, the Colorado river provides sand and silt. In the deeper basins, diatom and radiolaria ooze is deposited (*Van Andel,* 1964).

The Izmir-Bursa segment of the Middle Anatolian ophiolite zone is a fossil counterpart of the Gulf of California. During the Cenozoic, the northern part of the area became uplifted and more eroded than the southern part. Therefore, the bottom of the ophiolite trough is exposed north of Balıkesir; it is composed of ultramafitite massifs surrounded by crystalline schist. South of Balıkesir, the content of the trough is preserved. It consists of:

	South West of Izmir	Between Izmir and Manisa	*North* Near Akhisar
Late Maestrichtian and ? Paleocene	Sandstone flysch prevalent, few radiolarites and volcanites	Sandstone flysch and volcanites prevalent, few radiolarites	Volcanites and radiolarites prevalent, sandstone flysch and limestone decreasing
Early Maestrichtian to Cenomanian	------------	Neritic limestone	

The facies differences can be understood if the ophiolite zone is regarded as a trough which deepened toward the north and became filled with water from the same direction. At first, it only reached as far as Manisa, and limestone formed in its shallowest portion. During the late Cretaceous it extended as far as Izmir. From a former Aegean mainland rivers contributed clastics which were deposited as flysch. Near Akhisar, during the entire Late Cretaceous period, radiolarites were alternating with the products of a basaltic volcanism.

3. Strike-slip faults, accompanied by dilation, are also known on the ocean floor. The Yucatan Basin and the Cayman Trough in the Caribbean Sea are narrow and elongated deep-sea troughs formed in this way. They possess a disrupted, occasionally steeply sloping bottom of ultramafitic rocks (*Eggler* et al., 1973; *Dillon* and *Vedder,* 1973). Also the fracture zones described by *Thompson* and *Melson* (1972) are topographically and petrographically similar to the ophiolite troughs. Both are characterized by the association of ultramafitite with alkali-magmatites differing from the tholeiitic volcanism of the mid-ocean ridges.

4. There are similarities between the ophiolite zones and the aulacogenes which are also considered to be dilational structures. On the surface, they appear as grabens; in their subsurface, the crust is thinned out and the Moho elevated. Their fill is partly composed of sediments, such as in the Donez trough (*Chekunov,* 1967) and in the Baikal rift zone (*Florensov,* 1969; *Artemjev* and *Artyushkov,* 1971), and partly of magmatites, such as in the Oslo graben (*Ramberg* and *Smithson,* 1971) and in the Lake Superior rift (*Chase* and *Gilmer,* 1973).

These considerations suggest that the rift hypothesis fits the Anatolian ophiolite zones better than the explanation by plate tectonics. The ophiolite zones can be regarded as evidence for rifting of the Anatolian crust, opening never exceeded more than 50 to 200 km. The following geologic processes can be anticipated:

During the early Mesozoic, several zones of weakening formed at the bottom of the Tethys, caused by wrench faulting and dilation. In places, these zones followed the then existing Alpidic geosynclines, and in places they were independent. At the beginning, the zones may only have differed from their surroundings by greater depth and submarine basaltic eruptions. The aulacogen stage was soon replaced by the rift stage. Ultramafitite blocks pearced the crust and emplaced in the weak zones. They may have risen vertically, as in the case study by *Bonini* et al. (1973). But inclined fault planes also have to be considered. Overthrusted ultramafitite blocks have mostly been described from continental margins (*Milsom,* 1973; *Allemann* and *Peters,* 1972). But they also occur within the continents, such as at Ivrea. In Anatolia, sporadically intrusions of granite into ophiolite zones occur (*Chaput,* 1939; *Ketin,* 1947a: 265; *Holzer* and *Colin,* 1957: 225; *Lisenbee,* 1971: 366). The simplest explanation is to assume sialic rocks underlying the ultramafic ones. After emplacement of the ultramafitite blocks, an isostatic equilibration took place. The zone with ultramafitite massifs started to subside and turned into a eugeosynclinal deep-sea trough. Slumping and resedimentation are evidence of its tectonic instability. The submarine volcanism may have been stimulated by the rise of the ultramafitite. The conditions under which parts of the ophiolite sequences turned into blueschist are unknown. The termination of the eugeosynclinal stage was caused either by a gradual shallowing of the trough or by a tectogenesis which turned its contents into a „mélange".

The Holocene Tectonic Stress Conditions (Fig. 67, 68)

Insight into the present tectonic condition within the earth's crust may be gained from geophysical and geological investigations. On the one hand, movements can be deduced from earthquake focal mechanisms. On the other hand, the rules of geotectonics allow to extrapolate the structural development of an area from its geological past into the present. With reference to *Nowroozi* (1972) and *McKenzie* (1972), the following areas of young crustal movements are described here, first by themselves and subsequently in their relation to one another.

1. At the beginning of the Quaternary, West Anatolia, the western Taurus, and the basement of the Aegean Sea were turned into a block-faulted area (p. 94). It is dissected by numerous normal faults without regard for the older structures. From the western to the southern coast, the strike of the faults turns from west to south. The earthquake mechanisms are in accordance with the macrotectonics. From Adapazarı in the north to Muğla in the south, expansion perpendicular to the faults is predominant.

2. The North Anatolian seismic line and wrench fault obviously consists of segments of different age. Its middle part between Adapazarı and Erzincan may already have originated in Mesozoic time. Here, the line runs parallel to an ophiolite zone and is accompanied by a chain of Tertiary basins whose fillings go back as far as the Pliocene, locally even as far as the Eocene (p. 100). In northwestern Anatolia, west of Adapazari, the seismic line splits up into three branches. The two southern branches, Bursa-Gönen and Iznik-Bandırma, were mainly active during the early Quaternary period, as manifested by the considerable vertical displacements of the Thracian peneplain. The northernmost branch, i. e. Izmit-Tekirdag-Gulf of Saros, is lined by young unfilled grabens, and in the northern Aegean Sea perhaps by an active structure (p. 110). In eastern Anatolia, east of Erzincan, the ophiolite zone and the chain of basins are separating from the main seismic line. The former continue eastward beyond Erzurum. The deposits in the basins begin with the lower Pliocene; they are dissected by longitudinal faults (*Irrlitz,* 1972: 61). The seismic line turns from Erzincan towards the Van Gölü in the south-

east. Along this segment no older structures exist. Probably the Erzurum branch was seismically active till the end of the Tertiary, whereas the Van Gölü branch represents a young extension of the system. All earthquakes along the line are caused by strike-slip movements on planes with a strike of 90 to 105°. From Adapazarı to beyond Erzincan, the centers show a strict linear arrangement; in the northern Aegean region and the environs of the Van Gölü, however, they are scattered over larger areas.

3. The Malatya-Bitlis nappe in Southeast Anatolia as well as its continuation, the Zagros thrust zone, underwent two tectogenic phases. In Southeast Anatolia, the first thrusting occurred between the Campanian and the early Eocene (p. 89), in Iran it took place during the same period (*Braud,* 1970) or between the Turonian and Maestrichtian (*Ricou,* 1973), respectively. In Anatolia, the thrust was more than 15 km wide, and in the Zagros 40 to 50 km. After a period of quiescence from the Middle Eocene to the end of the Miocene, the tectogenesis revived during Pliocene time and may have lasted until the Quaternary. The Bitlis nappe was again thrusted by more than 15 km (p. 75) and the Zagros nappe by more than 40 km (*Braud,* 1973). The Southeast Anatolian part of the Iranides is at present tectonically inactive. In Iran, a steep, straight fault has replaced the inclined nappe plane (*Braud* and *Ricou,* 1971). According to seismic observations, southwest-vergent upthrusting with a right-lateral horizontal component is active along this line (*Canitez,* 1969: 81; *Tchalenko* et al., 1974).
4. The East Anatolian seismic line (p. 100) did not show strong movements until recent time. No precise information about the earthquake mechanism is available.
5. The Hatay-Jordan line represents like the East Anatolian line a left-lateral strike-slip fault. The former, however, is considerably older than the latter. The Hatay-Jordan line was the scene of two phases of considerable displacement, one by 60 km at the beginning and one by 45 km at the end of the Tertiary period (*Freund* et al., 1970: 124).
6. The Ecemis wrench fault, which in the eastern Taurus displaced the Ala Dağ by 75 km to the north as compared to the Bolkar Dağ, coincides with the Jordan line in strike and direction of movement. It also originated during the early Tertiary (p. 89).
7. In the northern Levantine Sea, the Hellenic trough follows the outcrop of a Benioff zone which dips toward the Aegean Sea (p. 111); an observation that is confirmed by seismic data. The overthrusting of the Aegean plate upon the Levantine plate progresses in a southwesterly direction near the western Peloponnese and in a southern direction near Crete. In the eastern part of the arc, near Rhodes, the compression is replaced by strike-slip movements.

The above-described phenomena can be brought into a larger geologic frame. Starting with the latest Cretaceous epoch, the Syrian-Arabian plate began to drift north along the Jordan line with a slight rotation to the left. At its northern and northeastern edge it thus created compressional structures, and at its southern and southwestern edge it left a dilatational area (*Wilson,* 1965). So, the opening of the Red Sea is complementary to the folding of the Iranides with regard to age, amount, and direction of crustal movements. The Red Sea rift originated during the early Tertiary. As to the exact moment no information is available, but it is possible that it coincided with the Laramian folding of the Iranides and the first displacement along the Hatay-Jordan line. The forming of the Ecemiş fault shows, that a considerable part of Anatolia was subjected to the same stress. The second phase of the opening of the Red Sea occurred 3 to 5 m. y. ago (*Girdler* and *Styles,* 1974). It was accompanied by renewed tectogenic events in the Iranides and the Jordan line.

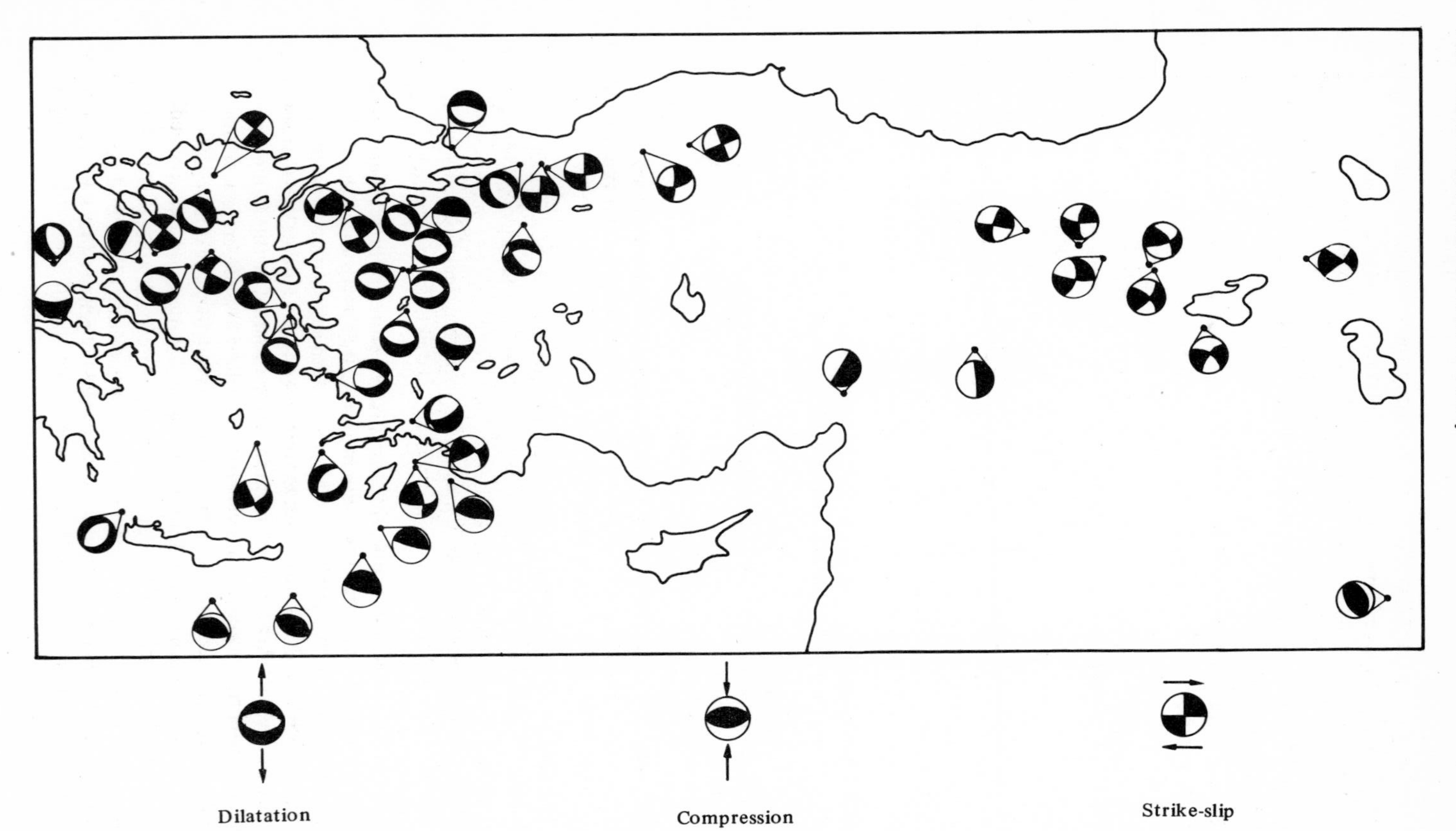

Fig. 67 Focal mechanisms of earthquakes occurring in the Eastern Mediterranean and Near East regions during the period 1936–69 (after *McKenzie*, 1973)

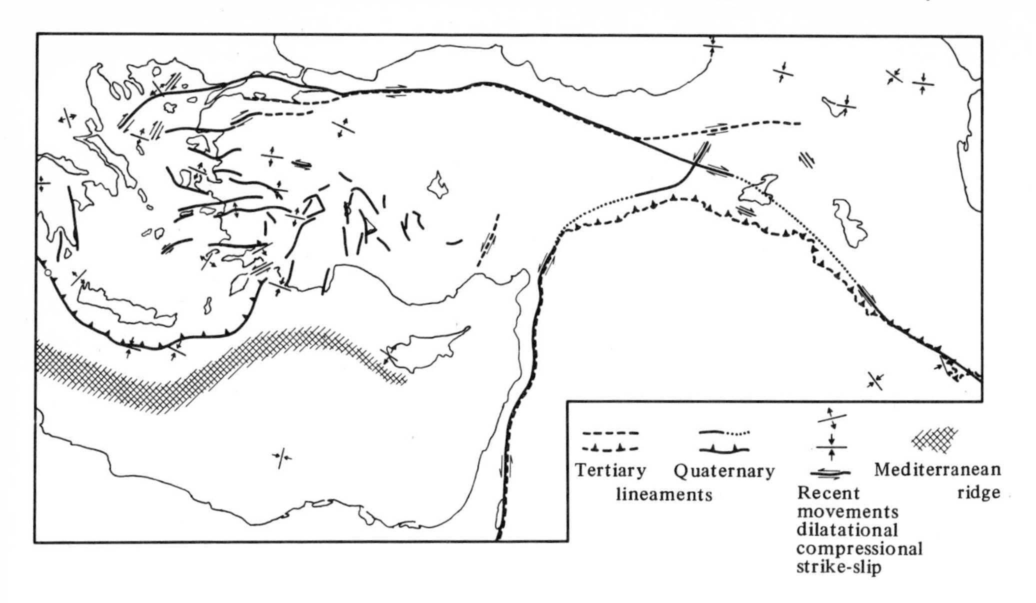

Fig. 68 Some Tertiary and Quaternary seismo-tectonic lineaments in the Eastern Mediterranean and Near East regions (after *Ketin,* 1969; *Becker-Platen,* 1970; *Bering,* 1971; *Irrlitz,* 1972; *McKenzie,* 1973, and others)
○ west of Crete = Upthrust of Cretaceous on Pliocene rocks in JOIDES station 127

A change took place at the turn of the Tertiary. The drifting of the Syrian-Arabian plate to the north continued, but thrusting and folding of the Iranides came to a standstill. From that period on, tensions had to be released in a different manner. A pair of conjugate wrench faults that cross one another northwest of the Van Gölü, took over the kinematic role of the Iranide thrusts and folds. One of these shear planes is northwest-striking. Its East Anatolian portion forms a young, seismically active branch of the North Anatolian earthquake line, pointing from Erzincan towards the Van Gölü. In Iran, it is represented by the Zagros zone. Near the Iranian-Iraqi border, the earthquake line departs from the overthrust, and obviously tends to connect with the North Anatolian line via Van Gölü. The second, northeast-striking shear plane is likewise not entirely developed. Portions of it, the East Anatolian line and the Hatay-Jordan line, are distinct. Like the Van Gölü region, the intermediate area between Elaziğ and Maraş shows an increased earthquake frequency, but no clear-cut fault trace.

On both shear planes the pressure from the Syrian-Arabian plate was transmitted to the Anatolian plate. The latter shifted to the west, and in the Aegean region to the southwest, parallel to the directional change of the North Anatolian line. In the arc of Crete it meets the African plate and is about to override it in a thrust plane that reaches down to the mantle. The reactions to this shift become evident in the neotectonics of the Eastern Mediterranean region; they are different in either plate. The Anatolian plate was brittle. In the direction of the maximal compressional stress, a bundle of normal faults is radiating from the interior of the West Anatolian mainland toward the coasts. The margin of the African plate was deformed by warping. The doming of the Mediterranean ridge presumably introduces the formation of foreland folds.

References

Please see foreword and chapter 2 for older publications.

Abbreviations:

Bull. MTA	=	Bulletin of the Mineral Research and Exploration Institute of Turkey (MTA Derg)
Publ. MTA	=	Publications de l'Institut d'Etudes et de Recherches Minières
Mecm. MTA	=	Maden Tetkik ve Arama Enstitüsü Mecmuasi
Bull. GST	=	Bulletin of the Geological Society of Turkey
Rev. Ist.	=	Revue de la Faculté des Sciences de l'Université d'Istanbul (Sciences Naturelles)
Mon. Ist.	=	Istanbul Üniversitesi Fen Fakültesi Monografileri (Tabii Ilimler kismi)

Abbate, E., V. Bortolotti and *P. Passerini:* Paleogeographic and tectonic considerations on the ultramafic belts in the Mediterranean area. Boll. Soc. Geol. It., Roma, 91 (1972), 239–282.
Abdüsselâmoğlu, S.: Almacikdağı ile Mudurnu ve Göynük civarinin jeolojisi. Ist. Mon. 14 (1959), 94 p.
Abdüsselâmoğlu, S.: Nouvelles observations stratigraphiques et paléontologiques dans les terrains paléozoiques affleurant à l'est du Bosphore. Bull. MTA 60 (1963), 1–6.
Acar, A.: Tortum bölgesinin jeologik etüdü. Doc. Tezi Univ. Erzurum (1970), 37 p.
Adams, F. D.: The birth and development of the Geological Sciences, New York, 2nd ed. (1954), 508 p.
Afshar, F. A.: Geology of Tunceli – Bingöl region of eastern Turkey. Bull. MTA 65 (1965), 33–44.
Agralı, B., E. Akyol and *Y. Konyalı:* Preuves palynologiques de l'existence du Dogger dans la région de Bayburt. Bull. MTA 65 (1965), 45–57.
Agralı, B. and *E. Akyol:* Étude palynologique des charbons de Hazro et considérations sur l'âge des horizons lacustres du Permo-Carbonifère. Bull. MTA 68 (1967), 1–26.
Agralı, B. and *Y. Konyalı:* Étude des microspores du bassin Carbonifère d'Amasra. Bull. MTA 73 (1969), 45–128.
Akarsu, I.: Geology of Mut region. Bull. MTA 54 (1960), 38–43.
Akarsu, I.: Geology of the Egean region (Babadağ and adjacent area). Bull. GST 12 (1969), 1–9.
Akartuna, M.: On the geology of Izmir–Torbalı–Seferihisar–Urla district. Bull. MTA 59 (1962a), 1–18.
Akartuna, M.: Çaycuma–Devrek–Yenice–Kozcağız bölgesinin jeolojisi. Ist. Mon. 17 (1962b), 58 p.
Akartuna, M.: Die Fortsetzung der Überschiebung von Şile an der Nordküste des Bosporus. Bull. MTA 61 (1963), 15–21.
Akartuna, M.: Armutlu yarımadasının jeolojisi. Ist. Mon. 20 (1968), 105 p.
Akkus, M.: Geologic and stratigraphic investigation of the Darende–Balaban basin (Malatya, ESE-Turkey). Bull. MTA 76 (1971), 1–54.
Akyol, E.: Contribution à l'étude palynologique des charbons tertiaires de la Turquie. Bull. MTA 63 (1964), 33–46.
Allasinaz, A., M. Gutnic and *A. Poisson:* La formation de l'Isparta Cay; calcaires à halobies, grès à plantes et radiolarites d'âge Carnien-Norien (Taurides, région d'Isparta, Turquie). In: Zapfe, H. (Ed.): Die Stratigraphie der alpin-mediterranen Trias, Wien 1974, 11–21.
Allemann, F. and *T. Peters:* The ophiolite-radiolarite belt of the North Oman Mountains. Ecl. Geol. Helv., Basel, 65 (1972), 657–697.
Allen, C. R.: Geological criteria for evaluating seismicity. Bull. Geol. Soc. Am., Boulder, 86 (1975), 1041–1057.
Alp, D.: Amasya yöresinin jeolojisi. Ist. Mon. 22 (1972), 101 p.
Alpan, S. and *G. Lüttig:* The German-Turkish lignite exploration in Turkey of the years 1965 to 1969. Newsl. Stratigr., Leiden, I (3) (1971), 11–18.
Altınlı, E.: Étude géologique de la chaîne côtière entre Bandırma et Gemlik. Rev. Ist. 8 (1943), 76–137.
Altınlı, E.: Étude stratigraphique de la région d'Antalya. Rev. Ist. 9 (1944), 227–238.
Altınlı, E.: The geology of the western portion of the Filyos river. Rev. Ist. 16 (1951a), 153–188.
Altınlı, E.: Geology of the Kayışdağı region. Rev. Ist. 16 (1951b), 189–205.
Altınlı, E.: Geology of the Iliksu region (Zonguldak, Turkey). Rev. Ist. 16 (1951c), 301–323.
Altinlı, E.: Geology of Siirt. Rev. Ist. 19 (1954a), 1–30.
Altınlı, E.: Geology of Hakkâri. Rev. Ist. 19 (1954b), 33–65.
Altınlı, E.: Are the Çamlıcas an overthrust sheet? Rev. Ist. 19 (1954c), 213–222.
Altınlı, E.: The geology of southern Denizli. Rev. Ist. 20 (1955), 1–47.
Altınlı, E.: Geology of the western portion of Pelitovasi. Rev. Ist. 21 (1956), 9–25.
Altınlı, E.: Geologic and hydrologic investigation of the Inegöl basin. Rev. Ist. 28 (1963), 173–199.
Altınlı, E.: Geology of eastern and southeastern Anatolia. Bull. MTA 66 (1966), 35–76, 67 (1966), 1–22.
Altınlı, E.: Geologic investigation of the Izmit–Hereke–Kurucadağ area. Bull. MTA 71 (1968), 1–28.
Altınlı, E., E. Demirtaşlı, E. B. Fritz, A. Hagshenow, M. W. A. Iqbal, H. Rahman, M. Samini and *N. N. Tilev:* Correlation of Cretaceous system in Turkey, Iran and Pakistan. CENTO, Rep. 2. Stratigr. Work. Group, Ankara, 1969, 83 p.
Altınlı, E., N. Söytürk and *K. Saka:* Geology of the Hereke–Tavşancıl–Tepecik area. Rev. Ist. 35 (1970a), 69–75.
Altınlı, E., O. Gürpınar and *S. Erşen:* Geology of the Erenköy–Deresakarı area (Bilecik province). Rev. Ist. 35 (1970b), 77–83.
Altınlı, E. and *C. Yetiş:* Geologic investigation of the Bayırköy–Osmaneli (Bilecik) area. Rev. Ist. 37 (1972), 1–18.

Ambraseys, N. N. and *J. S. Tschalenko:* Seismotectonic aspects of the Gediz, Turkey, earthquake of March 1970. Geophys. Journ., Oxford, 30 (1972), 229–252.

Angel, F.: Aus der Gesteinswelt Anatoliens. N. Jb. Min. Geol. Pal., B.-Bd. 62 A (1931), 57–162.

Anonymous: Karadeniz jeofizik araştırmalar, I. Jeomagnetik. Deniz Kuvv. Kom. Hidrogr. Neşr., Istanbul-Çubuklu., (1965), 6 p.

Anonymous: Symposium on the North Anatolian fault and earthquake belt. MTA, Ankara, 1973, 170 p.

Ardos, M.: Problèmes géomorphologiques du versant sud du Taurus Central (Turquie méridionale). Méditerranée, Nizza, 1969, 233–256.

Argyriadis, I.: Le Paléozoique supérieur métamorphique du massif d'Alanya (Turquie méridionale). Bull. Soc. Géol. Fr., Paris, VII, 16 (1974), 112–115.

Arni, P.: Zur Stratigraphie und Tektonik der Kreideschichten östlich Ereğli an der Schwarzmeerküste. Ecl. Geol. Helv., Basel, 24 (1931), 305–345.

Arni, P.: Zum Erdbeben zwischen Kırşehir, Keskin und Yerköy. Publ. MTA B 1 (1938), 58 p.

Arni, P.: Tektonische Grundzüge Ostanatoliens und benachbarter Gebiete. Publ. MTA B 4 (1939a), 89 p.

Arni, P.: Neue geologische Gesichtspunkte für den Bergbau im westlichen Steinkohlenbecken Nordanatoliens. MTA Mecm. 4 (4) (1939b), 46–63.

Arni, P.: Geologische Beobachtungen in den südlichen Ketten der Bitlis-Berge im Abschnitt des Başor-Cay westlich Siirt (Südostanatolien). MTA Mecm. 5 (4/21) (1940), 527–558.

Arni, P.: Materialien zur Altersfrage der Ophiolithe Anatoliens. MTA Mecm. 7 (3/28) (1942), 472–488.

Arpat, E. and *E. Bingöl:* The rift system of western Turkey; thoughts on its development. Bull. MTA 73 (1969), 1–9.

Arpat, E. and *N. Özgül:* Rock glaciers in Geyik Dağı area, Central Taurus. Bull. MTA 78 (1972), 28–32.

Arpat, E. and *F. Şaroglu:* The East Anatolian fault system; thoughts on its development. Bull. MTA 78 (1972), 33–39.

Artemjev, M. E. and *E. V. Artyushkov:* Structure and isostasy of the Baikal Rift and the mechanism of rifting. Journ. Geophys. Res., Washington D. C., 76 (1971), 1197–1211.

Artüz, S.: Amasra–Tarlaağzı kömür bölgesindeki kalin ve aradamarların (Westfalien C) mikrosporolojik etüdü ve korelasyon denemesi. Ist. Mon. 19 (1963), 70 p.

Artüz, S.: Zonguldak–Kozlu bölümdeki (Westfaliyen A) Hacıpetro kömür damarının petrografi incelemesi. Ist. Mon. 24 (1974), 32 p.

Aslaner, M.: Étude géologique et pétrographique de la région d'Edremit–Havran (Turquie). Publ. MTA 119 (1965), 98 p.

Aslaner, M.: Iskenderun–Kırıkhan bölgesindeki ofiyolitlerin jeoloji ve petrografisi. Publ. MTA 150 (1973), 78 p.

Assereto, R.: Notes on the Anisian biostratigraphy of the Gebze area (Kocaeli Peninsula, Turkey). Zeitschr. Deutsch. Geol. Ges., Hannover, 123 (1972), 435–444.

Assereto, R.: Aegean and Bithynian; proposal for two new Anisian substages. In: Zapfe, H. (Ed.): Die Stratigraphie der alpin-mediterranen Trias, Wien 1974, 23–39.

Astre, G. and *F. Charles:* Note sur les petites Toucasia d'Anatolie. Bull. Soc. Géol., Paris, V, 1 (1931), 697–705.

Ataman, G.: L'âge radiométrique du massif granodioritique d'Orhaneli. Bull. GST 15 (1972), 125–130.

Ataman, G.: L'âge radiométrique du massif granodioritique de Gürgenyayla. Bull. GST 16 (1) (1973), 22–26.

Atan, O. R.: Eğribucak–Karacaören (Hassa)–Ceylânlı–Dazevleri (Kırıkhan) arasındaki Amanos Dağlarının jeolojisi. Publ. MTA 139 (1969), 85 p.

Ayan, M.: Contribution à l'étude pétrographique et géologique de la région située au nord-est de Kaman (Turquie). Publ. MTA 115 (1963), 332 p.

Ayan, M.: Migmatites in the Gördes area. Bull. MTA 81 (1973), 85–109.

Ayan, M.: Gördes migmatitleri. MTA Derg. 81 (1973), 132–154.

Ayan, T. and *C. Bulut:* General geology of the area defined by the polygone Balaban–Yazıhan–Kurşunlu–Levent (vilayet Malatya). Bull. MTA 62 (1964), 60–73.

Aygen, T.: Étude géologique de la région de Balya. Publ. MTA D 11 (1956), 95 p.

Aykulu, A.: A geological investigation of an area to the south-east of Palu in south-eastern Turkey with special reference to the mineralization and economic potentialities. Thesis Fac. Sc. Univ. Leicester (1971), 196 p.

Aykulu, A. and *A. M. Evans:* Structures in the Iranides of south-eastern Turkey. Geol. Rundsch., Stuttgart, 63 (1974), 292–305.

Ayranci, B. and *M. Weibel:* Zum Chemismus der Ignimbrite des Erciyes-Vulkans (Zentral-Anatolien). Schweiz. Min.-Petr. Mitt., Zürich, 53 (1973), 49–60.

Bailey, E. B. and *W. J. McCallien:* Serpentinite lavas, the Ankara Mélange and the Anatolian thrust. Transact. R. Soc., Edinburgh, 62 (1954), 403–442.

Bailey, E. H. and *M. C. Blake Jr.:* Major chemical characteristics of Mesozoic Coast Range ophiolite in California. Jour. Research U. S. Geol. Surv., Reston, Va., 2 (1974), 637–656.

Barnaby, W.: Pleistocene footprint, Demirköprü, Salihli, Turkey. Nat., London, 254 (1975), 553.
Basarir, E.: The petrology and geology of the eastern flank of the Menderes massive on the east of Lake Bafa. Scient. Rep. Fac. Sc. Ege Univ., Bornova-Izmir, 102 (1970), 44 p.
Batum, I.: 1:25 000 ölcekli Adapazarı G 26–b3–nolu paftasın jeolojisi. MTA-Rep. 4778 (1968), 17 p. (unpublished).
Batum, I.: Report on the geology of Turkish–Iranian border, part I. MTA-Rep. 4303 (1969), 38 p. (unpublished).
Baykal, F.: Géologie de la région de Şile, Kocaeli (Bithynie), Anatolie. Rev. Ist. 7 (1942), 166–233.
Baykal, F.: Étude géologique du Taurus entre Darende et Kayseri (Anatolie). Rev. Ist. 10 (1945), 133–142.
Baykal, F.: Géologie de la région de Zile–Tokat–Yıldızeli. Rev. Ist. 12 (1947), 191–209.
Baykal, F.: Aperçu géologique des environs des montagnes de Şerafeddin et de Çotela (vilayets de Bingöl et de Diyarbakır, Anatolie orientale). Rev. Ist. 15 (1950), 134–152.
Baykal, F.: Lignes géologiques fondamentales de la région d'Oltu–Olur (NE de l'Anatolie). Rev. Ist. 16 (1951), 325–340.
Baykal, F.: Recherches géologiques dans la région de Kelkit–Şiran (NE de l'Anatolie). Rev. Ist. 17 (1952), 289–304.
Baykal, F.: Les terrains crétacés-tertiaires et les blocs exotiques entre Eflâni et Ulus (Anatolie NE). Rev. Ist. 19 (1954), 191–201.
Baykal, F.: Historik Jeoloji. Karadeniz Tekn. Üniv. Yayınl. 38, Istanbul (1971), 447 p.
Baykal, F. and *O. Kaya:* Note préliminaire sur le Silurien d'Istanbul. Bull. MTA 64 (1965), 1–8.
Baykal, F. and *Y. Tatar:* Erciyes volkanizmanın yaşı hakkında yeni gözlemler. Bull. GST 13 (2) (1970), 19–25.
Baykal, F. and *A. Kalafatçioğlu:* New geological observations in the area west of Antalya bay. Bull. MTA 80 (1973), 33–42.
Bayramgıl, O.: Sedimentpetrographische Untersuchungen im Steinkohlenbecken von Zonguldak (Türkei). Bull. GST 3 (1) (1951), 120–124.
Baysal, O.: New hydrous magnesium-borate minerals in Turkey: Kurnakovite, Inderite, Inderborite. Bull. MTA 80 (1973), 93–103.
Becker-Platen, J. D.: Lithostratigraphische Untersuchungen im Känozoikum Südwest-Anatoliens (Türkei). Beih. Geol. Jahrb., Hannover, 97 (1970), 244 p.
Becker-Platen, J. D. and *E. Löhnert:* Über Cardium-Funde aus dem Känozoikum der Umgebung von Söke (Westanatolien). Geol. Jahrb., Hannover, B 2 (1972), 55–63.
Beer, H.: Paläogeographie und Fazies der oberkretazischen südosttürkischen Phosphatprovinz. Bull. MTA 68 (1967), 84–88.
Belov, A. A.: Tectonic development of the alpine folded belt during the Paleozoic (Balkan Peninsula–Iranian Plateau–Pamirs). Geotectonics., Washington, 1967, 145–152.
Benda, L.: Grundzüge einer pollenanalytischen Gliederung des türkischen Jungtertiärs. Beih. Geol. Jahrb., Hannover, 113 (1971), 46 p.
Benda, L., F. Innocenti, R. Mazzuoli, F. Radicati and *P. Steffens:* Stratigraphic and radiometric data of the Neogene in Northwest Turkey. Zeitschr. Deutsch. Geol. Ges., Hannover, 125 (1974), 183–193.
Berger, E. H.: Geschichte der wissenschaftlichen Erdkunde der Griechen. Leipzig 1903, 666 p., 2nd ed.
Bering, D.: Lithostratigraphie, tektonische Entwicklung und Seengeschichte der neogenen und quartären Becken der Pisidischen Seenregion (Südanatolien). Beih. Geol. Jahrb., Hannover, 101 (1971), 150 p.
Bernard, J. H.: Mineralogy of the polymetallic ore deposits of Piraziz, vilayet Giresun, northeastern Turkey. Bull. MTA 75 (1970), 16–25.
Bernoulli, D., P. Ch. De Graciansky and *O. Monod:* The extension of the Lycian nappes (SW Turkey) into the southeastern Aegean islands. Ecl. Geol. Helv., Basel, 67 (1974), 39–90.
Berry, W. B. N. and *A. J. Boucot:* Correlation of the Southeast Asian and Near East Silurian rocks. Geol. Soc. Am. Spec. Pap., Boulder, 137 (1972), 65 p.
Beseme, P.: Le synclinal bitumineux de Kabalar (Göynük, Bolu). MTA-Rep. 4629 (1969), 26 p. (unpublished).
Besenecker, H., St. Dürr, G. Herget, V. Jacobshagen, G. Kauffmann, G. Lüdtke, W. Roth and *Kl.-W. Tietze:* Geologie von Chios (Ägäis). Geolog. et Palaeontolog., Marburg, 2 (1968), 121–150.
Besenecker, H. and *O. Otte:* Zur postalpidischen Sedimentation und Tektonik in der Ost-Ägäis. Zeitschr. Deutsch. Geol. Ges., Hannover, 123 (1972), 527–539.
Besenecker, H., F. Spitzenberger and *G. Storch:* Eine holozäne Kleinsäugerfauna von der Insel Chios, Ägäis. Senckenberg. Biol., Frankfurt, 53 (1972), 145–177.
Bilgütay, U.: Geology of the Hasanoğlan–Ankara region. Bull. MTA 54 (1960a), 44–51.
Bilgütay, U.: Some Permian calcareous algae from Ankara. Bull. MTA 54 (1960b), 52–65.
Bingöl, E.: Contribution à l'étude géologique de la partie centrale et sud-est du massif de Kazdağ (Turquie). Thèse Fac. Sc. Univ. Nancy, (1968), 190 p.
Bingöl, E.: Essaie d'application de mesures géochronologiques au massif de Kazdağ, Turquie. Bull. GST 14 (1971), 1–16.

Birot, P., L. Faugères, P. Gabert and *Y. Pechoux:* Esquisse géomorphologique de Sud-Ouest de l'Asie Mineure. Méditerranée, Gap, 9 (1968), 97–138.

Bittel, K.: Grundzüge der Vor- und Frühgeschichte Kleinasiens. Tübingen 1950, 135 p., 2nd ed.

Bittner, A.: Neue Brachiopoden und eine neue Halobia der Trias von Balia Maden. Jahrb. Geol. Reichsanst., Wien, 45 (1896), 249–254.

Blanckenhorn, M.: Das marine Miozän in Syrien. Denkschr. Ak. Wiss. Math.-nat. Kl., Wien, 57 (1890), 589–620.

Blümel, G.: Zur Stratigraphie des Küsten-Taurus bei Anamur (Türkei). Zeitschr. Deutsch. Geol. Ges., Hannover, 119 (1969), 426–442.

Blumenthal, M.: Un aperçu de la géologie du Taurus dans les vilayets de Niğde et Adana. Publ. MTA B 6 (1941), 95 p.

Blumenthal, M.: Contribution à la connaisance du Permo-Carbonifère du Taurus entre Kayseri-Malatya. MTA Mecm. 9 (1/31) (1944a), 105–133.

Blumenthal, M.: Schichtenfolge und Bau der Tauruskette im Hinterland von Bozkır (Vilayet Konya). Rev. Ist. 9 (1944b), 95–125.

Blumenthal, M.: Un gisement de bauxite dans le Permo-Carbonifère du Taurus oriental. MTA Mecm. 9 (2/32) (1944c), 218–225.

Blumenthal, M.: Sind gewisse Ophiolithzonen Nordanatoliens praeliassisch? MTA Mecm. 10 (1/33) (1945), 115–132.

Blumenthal, M.: Die neue geologische Karte der Türkei und einige ihrer stratigraphisch-tektonischen Grundzüge. Ecl. Geol. Helv., Basel, 39 (1946), 277–289.

Blumenthal, M.: Geologie der Taurusketten im Hinterland von Seydisehir und Beysehir. Publ. MTA D 2 (1947a), 242 p.

Blumenthal, M.: Das paläozoische Fenster von Belemedik und sein mesozoischer Kalkrahmen. Publ. MTA D 3 (1947b), 93 p.

Blumenthal, M.: Un apperçu de la géologie des chaînes nordanatoliennes entre l'ova de Bolu et le Kızılırmak inférieur. Publ. MTA B 13 (1948), 265 p.

Blumenthal, M.: Die Bauxitvorkommen der Berge um Akseki. Publ. MTA B 14 (1949), 59 p.

Blumenthal, M.: Beiträge zur Geologie der Landschaften am mittleren und unteren Yeşil Irmak (Tokat, Amasya, Havza, Erbaa, Niksar). Publ. MTA D 4 (1950), 153 p.

Blumenthal, M.: Recherches géologiques dans le Taurus occidental dans l'arrière-pays d'Alanya. Publ. MTA D 5 (1951), 134 p.

Blumenthal, M.: Das taurische Hochgebirge des Aladağ; neuere Forschungen zu seiner Geographie, Stratigraphie und Tektonik. Publ. MTA D 6 (1952a), 136 p.

Blumenthal, M.: Sur l'inconstance du déjettement tectonique dans la zone orogénique anatolienne. Rep. 18. Int. Geol. Congr., London, pt. 13 (1952b), 23–32.

Blumenthal, M.: Geologie des Hohen Bolkardağ, seiner nördlichen Randgebiete und westlichen Ausläufer (Südanatolischer Taurus). Publ. MTA D 7 (1955), 172 p.

Blumenthal, M.: Les chaînes bordières du Taurus au sud-ouest du bassin de Karaman–Konya et le problème stratigraphique de la formation schisto-radiolaritique. Bull. MTA 48 (1956), 1–39.

Blumenthal, M.: Der Vulkan Ararat und die Berge seiner Sedimentumrandung. Rev. Ist. 23 (1958), 177–327.

Blumenthal, M.: Le système structural du Taurus sud-anatolien. Livre Mém. *P. Fallot,* Paris, (1960/63), 611–662.

Blumenthal, M. and *G. Van Der Kaaden:* Catalogue of active volcanoes of the world including solfatara fields, part 17, Turkey. Roma 1964, 13 p.

Boccaletti, M., V. Bortolotti and *M. Sagri:* Ricerce sulle ofioliti delle catene alpine. I. Osservazione sull'Ankara Mélange nella zona di Ankara. Boll. Soc. Geol. It., Roma, 85 (1966a), 485–508.

Boccaletti, M., V. Bortolotti and *M. Sagri:* Ricerce sulle ofioliti delle catene alpine. 3. Arenarie ofiolitifere nella „Jurassic volcanic facies" a sud-ovest di Bolu (Zonguldak–Turchia). Boll. Soc. Geol. It., Roma, 85 (1966b), 525–528.

Boccaletti, M., V. Bortolotti, P. G. Malesani, P. Manetti, G. Papani and *F. P. Sassi:* Preliminary report on the geologic and petrographic mission in the Pontic Ranges (Turkey, summer 1968). Boll. Soc. Geol. It., Roma, 87 (1968), 667–676.

Böhm, J.: Beitrag zur Kenntnis der Senonfauna der bithynischen Halbinsel. Palaeontogr., Stuttgart, 69 (1927), 187–222.

Bojadjiev, St.: Tectonic structures of Bulgaria (Summary). In: *Yovchev, Y.* (Ed.): Tectonic structure of Bulgaria, Sofia, 1971, p. 467–558.

Bonini, W. E., T. P. Loomis and *J. D. Robertson:* Gravity anomalies, ultramafic intrusions and the tectonics of the region around the Strait of Gibraltar. Journ. Geophys. Res., Washington D. C., 78 (1973), 1372–1382.

Borsi, S., G. Ferrara, F. Innocenti and *R. Mazzuoli:* Geochronology and petrology of recent volcanics in the eastern Aegean Sea (West Anatolia and Lesvos Island). Bull. Volcanol., Napoli, 36 (1973), 473–496.

Bortolotti, V. and *M. Sagri:* Ricerce sulle ofioliti delle catene alpine. 4. Osservazioni sull'età e la giacitura delle ofioliti tra Smirne ed Erzurum (Turchia). Boll. Soc. Geol. It., Roma, 87 (1968), 661–666.

Bourgoin, A.: Sur les anomalies de la pesanteur en Syrie et au Liban. Notes Mém. Syrie–Liban, Paris, 4 (1945/48), 59–90.

Braud, J.: Les formations du Zagros dans la région de Kermanshah (Iran) et leurs rapports structuraux. C. R. Ac. Sc., Paris, 271 (1970), 1241–1244.

Braud, J.: La nappe de Kuh-e-Garun (région de Kermanshah, Iran), chevauchement de l'Iran central sur le Zagros. Bull. Soc. Géol. Fr., Paris, VII, 13 (1973), 416–419.

Braud, J. and *L.-E. Ricou:* L'accident du Zagros ou Main Thrust, un charriage et un coulissement. C. R. Ac. Sc., Paris, 272 (1971), 203–206.

Bremer, H.: Kleinasien. In: *Hölder, H.*: Jura, p. 488–492. Hdb. Stratigr. Geol. 4, Stuttgart 1964.

Bremer, H.: Zur Ammonitenfauna und Stratigraphie des unteren Lias (Sinemurium bis Carixium) in der Umgebung von Ankara (Türkei). N. Jahrb. Geol. Pal., Stuttgart, Abh. 122 (1965), 127–221.

Bremer, H.: Ammoniten aus dem unteren Bajocium und unteren Bathonium in der Umgebung von Ankara (Türkei). N. Jahrb. Geol. Pal., Stuttgart, Abh. 125 (1966), 155–169.

Brinkmann, R.: Die Südflanke des Menderes-Massivs bei Milas, Bodrum und Ören. Scient. Rep. Fac. Ec. Ege Univ., Bornova-Izmir, 43 (1967), 12 p.

Brinkmann, R.: Einige geologische Leitlinien von Anatolien. Geolog. et Palaeontolog., Marburg, 2 (1968), 111–119.

Brinkmann, R.: Das kristalline Grundgebirge von Anatolien. Geol. Rundsch., Stuttgart, 60 (1971a), 886–899.

Brinkmann, R.: The geology of western Anatolia. In: *Campbell, A. S.* (Ed.): Geology and History of Turkey, Tripoli, 1971b, p. 171–190.

Brinkmann, R.: Jungpaläozoikum und älteres Mesozoikum in Nordwest-Anatolien. Bull. MTA 76 (1971c), 55–67.

Brinkmann, R.: Mesozoic troughs and crustal structure in Anatolia. Bull. Geol. Soc. Am., Boulder, 83 (1972), 819–826.

Brinkmann, R.: Geologic relations between Black Sea and Anatolia. In: *Degens, E. T.* and *D. A. Ross* (Eds.): The Black Sea, geology, chemistry and biology. Am. Ass. Petr. Geol. Mem., Tulsa, 20 (1974), 63–76.

Brinkmann, R., R. Feist, W. U. Marr, E. Nickel, W. Schlimm and *H. R. Walter:* Geologie der Soma Daglari. Bull. MTA 74 (1970), 7–23.

Brinkmann, R., E. Flügel, V. Jacobshagen, H. Lechner, B. Rendel and *P. Trick:* Trias, Jura und Unterkreide der Halbinsel Karaburun (West-Anatolien). Geolog. et Palaeontolog., Marburg, 6 (1972), 139–150.

Brinkmann, R. and *R. Kühn:* Über Salzwasser-Thermen im Küstenland von West-Anatolien (Türkei). Chem. Geol., Amsterdam, 12 (1973), 171–187.

Brinkmann, R., H. Gümüs, F. Plumhoff and *A. A. Salah:* Höhere Oberkreide an der Küste West-Anatoliens und in Thrakien (to appear in 1976).

Brinkmann, R. and *O. Erol:* Geological bibliography of Turkey 1825–1975. Ankara (to appear in 1976).

Brönnimann, P., A. Poisson and *L. Zaninetti:* L'unité du Domuz Dağ (Taurus lycien, Turquie). Microfaciès et foraminifères du Trias et du Lias. Riv. It. Paleont., Milano, 76 (1970), 1–36.

Broili, F.: Geologische und paläontologische Resultate der Dr. Grothe'schen Vorderasienexpedition 1906 und 1907. In: *Grothe, H.*: Meine Vorderasienexpedition 1906 und 1907, Leipzig 1911, v. I, p. 1–70.

Brothers, B. N.: Lawsonite-Albite schists from northernmost New Caledonia. Contr. Min. Petr., Berlin–Heidelberg–New York, 25 (1970), 185–202.

Brown, W. W. and *K. D. Jones:* Borate deposits of Turkey. In: *Campbell, A. S.* (Ed.): Geology and History of Turkey, Tripoli 1971, p. 483–492.

Brunn, J. H.: Les zones helléniques internes et leur extension. Réflexions sur l'orogénèse alpine. Bull. Soc. Géol. Fr., Paris, VII, 2 (1960), 470–486.

Brunn, J. H.: Le problème de l'origine des nappes et de leurs translations dans les Taurides occidentales. Bull. Soc. Géol. Fr., Paris, VII, 16 (1974), 101–106.

Brunn, J. H., P.-Ch. De Graciansky, M. Gutnic, Th. Juteau, R. Lefèvre, J. Marcoux, O. Monod and *A. Poisson:* Structures majeures et corrélations stratigraphiques dans les Taurides occidentales. Bull. Soc. Géol. Fr., Paris, VII, 12 (1972a), 515–556.

Brunn, J. H., L. Faugères and *P. Robert:* Une nouvelle série du Jurassique moyen/Crétacé inférieur surmontant les ophiolithes dans le détroit de Kozani (Macédoine, Grèce). C. R. Soc. Géol. Fr., Paris 1972b, 26–27.

Brunnacker, K.: Affleurements de loess dans les régions nord-méditerranéennes. Rev. Géogr. Phys. Géol. Dyn., Paris, II, 11 (1969), 325–334.

Bürküt, Y.: Kuzeybati Anadoluda yeralan plütonlarin mukayeseli jenetik etüdü. Tez Ist. Tekn. Üniv. Maden Fak., Istanbul 1966, 272 p.

Buggisch, W.: Nachweis von Ludlow und Gedinne im Taurus (Südanatolien). N. Jahrb. Geol. Pal. Monatsh., Stuttgart, 1973, 264–272.

Buggisch, W., E. Flügel und *G.-F. Tietz:* Mitteldevonische Vulkanite im südanatolischen Taurus (Beiträge zur Biostratigraphie des anatolischen Paläozoikums 1). N. Jb. Geol. Pal. Monatsh., Stuttgart, 1974, 577–592.

Butterlin, J. and *O. Monod:* Biostratigraphie (Paléocène à Eocène moyen) d'une coupe dans le Taurus de Beysehir (Turquie). Ecl. Geol. Helv., Basel, 62 (1969), 583–604.

Caglar, K. Ö.: Mineral water and hot water springs in Turkey I–IV. Publ. MTA B 11, 107 (1947/50, 1961), 791 p.

Campbell, A. S. (Ed.): Geology and History of Turkey. Tripoli 1971, 511 p.

Canitez, N.: The focal mechanisms in Iran and their relations to tectonics. Pageoph., Basel, 75 (1969), 76–87.

Canuti, P., M. Marcucci and *C. Pirini Radrizzani:* Microfacies e microfaune nelle formazioni paleozoiche dell' anticlinale di Hazro (Anatolia sud-orientale, Turchia). Boll. Soc. Geol. It., Roma, 89 (1970), 21–40.

Chaput, E.: Voyages d'études géologiques et géomorphogénétiques en Turquie. Mém. Inst. Franç. d'Archéol. de Stamboul, Paris, 2 (1936), 312 p.

Chaput, E.: Observations sur les terrains crétacés et tertiaires du Taurus oriental. C. R. Ac. Sc., Paris, 208 (1939), 2091–2092.

Chaput, E. and *S. Gillet:* Les faunes des mollusques des terrains à Hipparion gracile de Küçük Çekmece près d'Istanbul (Turquie). Bull. Soc. Géol. Fr., Paris, V, 8 (1939), 363–388.

Chaput, G. and *B. Darkot:* Sur le Pliocène des environs d'Antalya. C. R. Ac. Sc., Paris, 236 (1953), 2259–2260.

Charles, F.: Note sur le Houillier d'Amasra (Asie Mineure). Ann. Soc. Géol. Belg., Liège, 54, B (1931), 151–178.

Charles, F.: Contribution à l'étude des terrains paléozoiques de l'Anatolie du Nord-Ouest (Asie Mineure). Mém. Soc. Géol. Belg., Liège, 1933, 53–152.

Charles, Fl. A.: Les actions tectoniques dans la région charbonnière du Nord de l'Anatolie et leur influence sur la sédimentation crétacée. C. R. 19. Int. Geol. Congr. fasc. 14, Alger (1954), 307–336.

Chase, C. G. and *T. H. Gilmer:* Precambrian plate tectonics: the Midcontinent gravity high. Earth Plan. Sc. L., Amsterdam, 21 (1973), 70–78.

Chazan, W.: Les gisements d'oxydes de manganèse de la région d'Héraclée (Karadeniz Ereğlisi) et leur genèse. MTA Mecm. 12 (1/37) (1947), 112–130.

Chekunov, A. V.: Mechanism responsible for structures of the aulacogen type (taking the Dniepr-Donets basin as an example). Geotectonics, Washington, 1967, 137–144.

Ciry, R.: Les Fusulinidés de Turquie. Ann. Paléont., Paris (1941/43) 29, 51–78, 30, 15–43.

Çoğulu, E.: Étude pétrographique de la région de Mihalıççık (Turquie). Schweiz. Min.-Petr. Mitt., Zürich, 47 (1967), 683–824.

Çoğulu, E.: Gümüşhane ve Rize granitik plütonlarının mukayeseli petrolojik ve jeokronometrik etüdü. Doc. Tezi Ist. Tekn. Üniv., Istanbul, 1970, 92 p.

Çoğulu, E., M. Delaloye and *R. Chessex:* Sur l'âge de quelques roches intrusives acides de la région d'Eskişehir (Turquie). Arch. Sc. Soc. Phys. Hist. Nat., Genève, 18 (1965), 692–699.

Çoğulu, E. and *D. Krummenacher:* Problèmes géochronométriques dans la partie NW de l'Anatolie Centrale (Turquie). Schweiz. Min.-Petr. Mitt., Zürich, 47 (1967), 825–831.

Colin, H. J.: Geologische Untersuchungen im Raume Fethiye–Antalya–Kaş–Finike. Bull. MTA 59 (1962), 19–61.

Collignon, M., S. Guérin-Franiatte, M. Gutnic and *Th. Juteau:* Découverte de Trias supérieur fossilifère à ammonites dans la région d'Eğridir (Sud-est du Taurus de Pisidie, Turquie). C. R. Ac. Sc., Paris, 270 (1970), 2244–2248.

Cordey, W. G.: Stratigraphy and sedimentation of the Cretaceous Mardin formation in southern Turkey. In: *Campbell, A. S.* (Ed.): Geology and History of Turkey, Tripoli 1971, p. 317–348.

Corsin, P. and *C. Martin:* Découverte d'un niveau à plantes dans un faciès marin du Malm, dans le Taurus occidental (Turquie). Ann. Soc. Géol. Nord, Lille, 89 (1969), 335–342.

Čtyroký, P.: Permian flora from the Ga'ara region (Western Iraq). N. Jahrb. Geol. Pal. Monatsh., Stuttgart, 1973, 383–388.

Cuif, J.-P. and *J.-Cl. Fischer:* Étude systématique sur les Chaetetida du Trias de Turquie. Ann. Paléont. (Invertebr.), Paris, 60 (1974), 1–14.

Daci-Dizer, A.: Contribution à l'étude paléontologique du Nummulitique de Kastamonu. Rev. Ist. 18 (1953), 207–299.

D'Archiac, A., P. Fischer and *E. De Verneuil:* Paléontologie. In: *De Tchihatcheff, P.:* Asie Mineure, Paris 1866, 442 p.

Daus, H.: Beiträge zur Kenntnis des marinen Miozäns in Kilikien und Nordsyrien. N. Jahrb. Min. Geol. Pal., Stuttgart, Beil.-Bd. 38 (1915), 429–500.
Davis, P. H.: Distribution patterns in Anatolia with particular reference to endemism. In: *Davis, P. H., P. C. Harper* and *J. C. Hedge* (Eds.): Plant life in South-West Asia. Edinburgh 1971, p. 15–27.
Dean, W. T.: The correlation and trilobite fauna of the Bedinan formation (Ordovician) in south-eastern Turkey. Bull. Brit. Mus. (Nat. Hist.), Geol., London, 15 (1967), 81–123.
Dean, W. T.: The Lower Paleozoic stratigraphy and faunas of the Taurus Mountains near Beyşehir, Turkey. II. The trilobites of the Seydişehir formation (Ordovician). Bull. Brit. Mus. (Nat. Hist.), Geol., London, 20 (1971), 1–24.
Dean, W. T.: The trilobite genus Holasaphus *Matthew* 1895 in the Middle Cambrian rocks of Nova Scotia and eastern Turkey. Canad. Journ. Earth Sc., Ottawa, 9 (1972), 266–279.
Dean, W. T.: The Lower Paleozoic stratigraphy and faunas of the Taurus Mountains near Beyşehir, Turkey III. The trilobites of the Sobova formation (Lower Ordovician). Bull. Brit. Mus. (Nat. Hist.) Geol., London, 24 (1973), 281–348.
Dean, W. T. and *R. Krummenacher:* Cambrian trilobites from the Amanos Mountains, Turkey. Paleontology, London, 4 (1961), 71–81.
Dean, W. T. and *O. Monod:* The Lower Paleozoic stratigraphy and faunas of the Taurus Mountains near Beyşehir, Turkey. I. Stratigraphy. Bull. Brit. Mus. (Nat. Hist.) Geol., London, 19 (1970), 411–426.
Degens, E. T. and *D. A. Ross* (Eds.): The Black Sea, geology, chemistry and biology. Am. Ass. Petr. Geol. Mem., Tulsa, 20 (1974), 633 p.
De Graciansky, P.: Le massif cristallin du Mendérès (Taurus occidental, Asie Mineure), un exemple possible du vieux socle granitique remobilisé. Rev. Géogr. Phys. Géol. Dyn., Paris, 8 (1966), 289–306.
De Graciansky, P.: Recherches géologiques dans le Taurus lycien. Thèse Univ. Paris-Sud, Paris, 1972, 762 p.
De Graciansky, P., Cl. Lorenz and *J. Magné:* Sur les étapes de la transgression du Miocène inférieur observé dans les fenêtres de Göcek (Sud-Ouest de la Turquie). Bull. Soc. Géol. Fr., Paris, VII, 12 (1972), 557–564.
Delaloye, M., E. Çoğulu and *R. Chessex:* Étude géochronométrique des massifs cristallins de Rize et de Gümüşhane, Pontides orientales (Turquie). C. R. Séanc. Soc. Phys. Hist. Nat., Genève, N. s. 7 (1972), 43–52.
Delépine, G.: Étude de quelques brachiopodes du Paléozoique des environs de Bartine-Zongouldak. Mém. Soc. Géol. Belg., Liège, 1933, 153–165.
Delpey, G.: Quelques gastéropodes maestrichtiens de Turquie. Bull. Soc. Géol. Fr., Paris, V, 8 (1938), 485–492.
Demircioğlu, A.: Boron minerals of Turkey: Hydroboracite. Bull. MTA 80 (1973), 104–117.
Demirtaşlı, E. and *C. Pisoni:* The geology of Ahlat–Adilcevaz area (North of Lake Van). Bull. MTA 64 (1965), 24–39.
Demirtaşlı, E., M. Selim, A. Z. Bilgin, N. Turan, S. Işıklar, D. Y. Sanlı and *F. Erenler:* Geology of the Bolkar Mountains. Preprint, Congr. Earth Sc., Ankara 1973.
De Peyronnet, Ph.: Esquisse géologique de la région d'Alanya (Taurus méridional), origine des bauxites métamorphiques. Bull. MTA 76 (1971), 90–116.
De Planhol, X.: Sur le réseau hydrographique et les dernières phases de soulèvement du Taurus occidental. C. R. Ac. Sc., Paris, 234 (1952), 1386–1388.
De Planhol, X.: Position stratigraphique et signification morphologique des travertins subtauriques de l'Anatolie sud-occidentale. Actes 4. Inqua-Congr., Roma, 2 (1956a), 467–471.
De Planhol, X.: Contribution à l'étude géomorphologique du Taurus occidental et de ses plaines bordières. Rev. Géogr. Alp., Grenoble, 44 (1956b), 609–685.
De Ridder, N. A.: Sediments of the Konya basin, Central Anatolia, Turkey. Palaeogeogr. Palaeoclim. Palaeoec., Amsterdam, 1 (1965), 225–254.
Desio, A.: Le isole italiane dell'Egeo. Mem. Carta Geol. d'Italia, Roma, 24 (1931), 546 p.
Desprairies, A. and *M. Gutnic:* Les grès rouges au sommet du Paléozoique et les niveaux ferrallitiques de la couverture mésozoique (Nord-Est du Taurus occidental, Turquie). Bull. Soc. Géol. Fr., Paris, VII, 12 (1972), 505–514.
Dessauvagie, T. F. J. and *Z. Dager:* Occurrences of Lasiodiscidae in Anatolia. Bull. MTA 60 (1963), 76–84.
De Tchihatcheff, P.: Asie Mineure. Description physique de cette contrée. 4. partie, Paléontologie et Géologie. Paris 1867/69, 2245 p.
Dewey, J. F., W. C. Pitman III, W. B. F. Ryan and *J. Bonnin:* Plate tectonics and the evolution of the Alpine system. Bull. Geol. Soc. Am., Boulder, 84 (1973), 3137–3180.
Dillon, W. P. and *J. G. Vedder:* Structure and development of the continental margin of British Honduras. Bull. Geol. Soc. Am., Boulder, 84 (1973), 2713–2732.
Dimitrijević, M. D. and *M. N.:* Olisthostrome mélange in the Yugoslavian Dinarides and late Mesozoic plate tectonics. Journ. Geol., Chicago, 81 (1973), 328–340.
Di Paola, G. M. and *F. Innocenti:* Rapport pétrographique à la suite de la mission en Anatolie. MTA Rap. 4726 (1969), 32 p. (unpublished).
Dizer, A.: Les foraminifères de l'Éocène inférieur de l'Ouest du ravin de Filyos. Rev. Ist. 21 (1956), 1–8.

Dizer, A.: Sur la faune des nummulites trouvés entre Akhisar et Sındırgı. Rev. Ist. 27 (1962a), 29–37.
Dizer, A.: Les foraminifères de l'Éocène et l'Oligocène de Denizli. Rev. Ist. 27 (1962b), 39–47.
Dizer, A.: La limite Crétacé/Tertiaire dans le bassin NW de la Turquie. Rev. Micropal., Paris, 14 (5) (1972), 43–47.
Dora, Ö.: Geologisch-lagerstättenkundliche Untersuchungen im Yamanlar-Gebirge nördlich von Karşıyaka (Westanatolien). Publ. MTA 116 (1964), 64 p.
Dora, Ö.: Petrologische und metallogenetische Untersuchungen im Granitmassiv von Karakoca. Bull. MTA 73 (1969), 10–27.
Dora, Ö.: Orthoklas-Mikroklin-Transformation in Migmatiten des Eğrigöz-Massivs. Bull. GST 15 (1972), 131–152.
Dubertret, L.: Données diverses sur le Pliocène et le Quaternaire marins de la Syrie et du Liban. Notes Mém. Syr. Liban, Paris, 2 (1937), 111–121.
Dubertret, L.: Géologie des roches vertes du nord-ouest de la Syrie et du Hatay (Turquie). Notes Mém. Moyen-Orient, Paris, 6 (1955), 1–224.
Dumitrashko, N. V. and *D. A. Lilienberg:* The recent tectonics of Caucasia. In: *Gerasimov, I. P.* et al. (Eds.): Recent crustal movements, Moskva, (1963), p. 291–302 (Engl. transl. Jerusalem 1967).
Dumont, J. F.: Découverte d'un horizon du Cambrien à trilobites dans l'autochthone du Taurus de Pisidie, région d'Eğridir, Turquie. C. R. Ac. Sc., Paris, D 274 (1972), 2435–2438.
Dumont, J. F., M. Gutnic, J. Marcoux, O. Monod and *A. Poisson:* Le Trias des Taurides occidentales (Turquie). Définition du bassin pamphylien: un nouveau domaine à ophiolithes à la marge externe de la chaîne taurique. Zeitschr. Deutsch. Geol. Ges., Hannover, 123 (1972), 385–409.
Durand, G. L. A.: Détermination de l'âge d'une pechblende turque, Dikmen (prov. de Muğla). Bull MTA 58 (1962), 145–146.
Durand-Delga, M. and *M. Gutnic:* Calpionelles du Taurus sud-anatolien (Turquie). C. R. Ac. Sc., Paris, 262 (1966), 1836–1839.
Durrani, S. A., H. A. Khan and *M. Taj:* Obsidian source identification by fission track analysis. Nat., London, 233 (1971), 242–245.

Egemen, R.: A preliminary note on fossiliferous Upper Silurian beds near Ereğli. Bull. GST 1 (1) (1947), 44–59.
Egemen, R.: On the significance of the flora found in the Ihsaniye beds at Kozlu, Zonguldak. Rev. Ist. 24 (1959), 1–24.
Egeran, N.: Contribution apportée aux connaissances sur la tectonique alpine par les études géologiques et tectoniques effectuées récemment en Turquie. MTA Mecm. 10 (2/34) (1945), 319–335.
Egeran, N.: Relations entre les unités tectoniques et les gîtes métallifères de Turquie. MTA Mecm. 11 (1/35) (1946), 40–49.
Egeran, N. and *E. Lahn:* Türkiye jeolojisi, Ankara (1948), 206 p.
Eggler, D. H., D. A. Fahlquist, W. E. Pequegnat and *I. M. Herndon:* Ultrabasic rocks from the Cayman trough, Caribbean Sea. Bull. Geol. Soc. Am., Boulder, 84 (1973), 2133–2138.
Eichholz, D. E. (Ed.): Theophrastos De Lapidibus.Oxford (1965), 141 p.
Eisma, D.: Beach ridges near Selçuk, Turkey. Tijdschr. K. Nederl. Aardrijksk. Gen., Amsterdam, II, 79 (1962), 234–246.
El Ishmawi, R.: Geologie des nördlichen Mittelteils des Amanosgebirges zwischen Islahiye und Bahce (S-Türkei). Geotekt. Forsch., Stuttgart, 42 (1972), 34–63.
Enay, R.: Faunes anatoliennes (Ammonitina, Jurassique) et domaines biogéographiques nord et sud téthysiennes. Bull. Soc. Géol. Fr., Paris 1975 (in press).
Enay, R., C. Martin, O. Monod and *J. P. Thieuloy:* Jurassique supérieur à ammonites (Kimméridgien-Tithonique) dans l'autochthone du Taurus de Beyşehir (Turquie méridionale). Ann. Inst. Geol. Hung., Budapest, 54 (2) (1971), 397–422.
Enderle, J.: Über eine anthracolithische Fauna von Balia Maden in Kleinasien. Beitr. Pal. Öst.-Ung., Wien, 13 (1901), 49–109.
Engelhardt, H.: Tertiärpflanzen aus Kleinasien. Beitr. Pal. Geol. Öst.-Ung., Wien, 15 (1903), 55–64.
Engin, T.: Petrology of the ultramafic rocks and brief geology of the Andızlık-Zımparalık area, Fethiye, Southwest Turkey. Bull. MTA 78 (1972), 1–18.
Erentöz, C.: Géologie du bassin de l'Aras. Bull. GST 5 (1954), 1–53.
Erentöz, C. and *Z. Ternek:* Les sources thermominérales de la Turquie et l'étude de l'énergie géothermique. Bull. MTA 70 (1968), 1–59.
Erentöz, L.: Stratigraphie des bassins néogènes de Turquie, plus spécialement d'Anatolie méridionale et comparaisons avec le domaine méditerranéen dans son ensemble. Publ. MTA C 3 (1956), 53 p.
Erentöz, L.: Mollusques du Néogène des bassins de Karaman, Adana et Hatay. Publ. MTA C 4 (1958), 196 p.
Erentöz, L. and *C. Öztemür:* Aperçu général sur la stratigraphie du Néogène de la Turquie et observations sur ses limites inférieure et supérieure. Curs. Conf. Inst. Mallada, Madrid, 9 (1964), 259–266.

Ergin, K., U. Güçlü and *Z. Uc:* Türkiye ve civarinin deprem katalogu. Ist. Tekn. Üniv. Maden Fak., Istanbul, (1967), 169 p.

Ergin, K., U. Güçlü and *G. Aksay:* A catalogue of earthquakes of Turkey and surrounding area (1965–1970). Ist. Tekn. Üniv. Arz Fiz. Enst. Yayınl., Istanbul, 28 (1971), 92 p.

Erinc, S.: Türkiye kuaterneri ve jeomorfolojinin katkısı. Jeomorfol. Derg., Ankara, 2 (2) (1970), 12–35.

Erk, S.: Étude géologique de la région entre Gemlik et Bursa. Publ. MTA B 9 (1942), 295 p.

Ernst, W. G.: Blueschist metamorphism and p-t-regimes in active subduction zones. Tectonophys., Amsterdam, 17 (1973), 255–272.

Erol, O.: Ankara ve civarinin jeolojisi hakkında rapor. MTA Rap. 2491 (1954), 238 p. (unpublished).

Erol, O.: Ankara güneydogusundaki Elma Dağı ve çevresinin jeoloji ve jeomorfolojisi üzerinde bir araştırma. Publ. MTA D 9 (1956), 99 p.

Erol, O.: 41/3–4, 42/3–4, 43/3 numeralı paftalar sahasının jeolojik reviziyon ve korelasyonu hk. rapor. MTA Rap. 2647 (1958), 50 p. (unpublished).

Erol, O.: The orogenic phases of the Ankara region. Bull. GST 7 (2) (1961), 75–85.

Erol, O.: Paleozoic formations and the problem of the Paleozoic/Mesozoic boundary in the Ankara region. Bull. GST 11 (1968), 1–20.

Erol, O.: Observations of Anatolian coastline changes during the Holocene. Çograf. Araşt. Derg., Ankara, 2 (1969), 89–102.

Erol, O.: Les hauts niveaux pleistocènes du Tuzgölü (Lac Salé) en Anatolie centrale (Turquie). Ann. Géogr., Paris, 79 (1970), 39–50.

Erol, O.: Geomorphological evidence of the recessional phases of the pluvial lakes in Konya, Tuzgölü and Burdur basins in Anatolia. Çograf. Araşt. Derg., Ankara, 3/4 (1972), 13–52.

Erol, O.: Geomorphological outlines of the Ankara area. Ank. Üniv. Dil ve Tarih-Çograf. Fak. Yayınl., Ankara, 240 (1973), 29 p.

Erol, O. and *C. P. Nuttall:* Some marine Quaternary deposits in the Dardanelles area (Turkey). Çograf. Araşt. Derg., Ankara, 5/6 (1973), 27–91.

Eroskay, S. O.: Geology of the Paşalar gorge, Gölpazarı area. Rev. Ist. 30 (1965), 135–170.

Flemming, N. C.: Eustatic and tectonic factors in the relative vertical displacement of the Aegean coast. In: *Stanley, D. J.* (Ed.): The Mediterranean Sea. Stroudsburg, Penn., (1972), p. 189–201.

Florensov, N. A.: Rifts of the Baikal mountain region. Tectonophys., Amsterdam, 8 (1969), 443–456.

Flügel, E.: Fazies-Interpretation der Cladocoropsis-Kalke (Malm) auf Karaburun (Westanatolien). Arch. Lagerst.-Forschg. d. Ostalpen, Sonderbd. 2, Leoben (1974), 79–84.

Flügel, E. und *H.:* Stromatoporen und Korallen aus dem Mitteldevon von Feke (Anti-Taurus). Senckenberg. Leth., Frankfurt, 42 (1961), 377–409.

Flügel, H.: Zur Paläontologie des anatolischen Paläozoikums. V. Graptolithen aus dem Gotlandium des Antitaurus. N. Jahrb. Geol. Pal. Monatsh., Stuttgart, 1955a, 478–488.

Flügel, H.: Mitteldevonfauna von Yahyalı (NO Ala Dağ, Taurus). N. Jahrb. Geol. Pal. Abh., Stuttgart, 101 (1955b), 267–280.

Flügel, H.: Bryozoen aus dem Perm des Ala Dağ. N. Jahrb. Geol. Pal. Abh., Stuttgart, 101 (1955c), 283–292.

Flügel, H.: Permische Korallen aus dem südanatolischen Taurus. N. Jahrb. Geol. Pal. Abh., Stuttgart, 101 (1955d), 293–318.

Flügel, H.: Die Entwicklung des vorderasiatischen Paläozoikums. Geotekt. Forsch., Stuttgart, 18 (1964), 68 p.

Flügel, H.: Paleozoic rocks of Turkey. In: *Campbell, A. S.* (Ed.): Geology and History of Turkey. Tripoli 1971, p. 211–224.

Flügel, H.: Einige Probleme des Variszikums in Neoeuropa. Geol. Rundsch., Stuttgart, 64 (1975), 1–62.

Flügel, H. and *E. Kıratlıoğlu:* Visékorallen aus dem Antitaurus. N. Jahrb. Geol. Pal. Monatsh., Stuttgart, 1956, 512–520.

Forbes, R. J.: Studies in ancient technology. v. 1–6, 2nd ed., v. 7–9, 1st ed., Leiden (1963/66), 2293 p.

Fourquin, C., I.-C. Paicheler and *J. Sauvage:* Premières données sur la stratigraphique du „Massif Galate“ d'andesite. C. R. Ac. Sc., Paris, 270 (1970), 2253–2255.

Fratschner, W.-Th.: Karbonfenster im Flysch Nordanatoliens. Roemeriana, Clausthal, 1 (1954), 209–226.

Fratschner, W.-Th. and *G. Van Der Kaaden:* Über die senone Effusiv-Mergel-Tuffit-Fazies im Raume Bartın–Kurucasile am Schwarzen Meer. Bull. GST 4 (1) (1953), 1–16.

Frech, F.: Geologische Beobachtungen im pontischen Gebirge. Oberkreide, Flysch und mitteltertiäre Masseneruptionen bei Trapezunt, Kerasunt und Ordu. N. Jahrb. Geol. Pal., Stuttgart, 1910, v. 1, 1–24.

Frech, F.: Geologie Kleinasiens im Bereich der Bagdadbahn. Zeitschr. Deutsch. Geol. Ges., Berlin, 68 (1916), 1–322.

Freneix, S.: Daonella indica (Bivalvia) de la région d'Antalya (bordure sud du Taurus de Turquie). Microstructure du test. Notes Mém. Moyen-Orient, Paris, 12 (1971/72), 173–181.

Freund, R., Z. Garfunkel, I. Zak, M. Goldberg, T. Weissbrod and *B. Derin:* The shear along the Dead Sea rift. Phil. Trans. R. Soc., London, A 267 (1970), 107–130.

Fuchs, Th.: Über einige Hieroglyphen und Fucoiden aus den paläozoischen Schichten von Hadjin in Kleinasien. Sitzber. Ak. Wiss. Math.-nat. Kl. pt. I, Wien, 111 (1902), 327–330.
Furon, R.: Introduction à la géologie et à l'hydrogéologie de la Turquie. Mém. Mus. Hist. Nat. sér. C, v. 3, fasc. 1, Paris (1953), 128 p.

Gabrielyan, A. H.: The tectonic structure of the Anticaucasus (the Minor Caucasus) and its position in the Mediterranean orogenic belt. Rep. 22. Int. Geol. Congr. pt. 11, New Delhi (1964), 400–419.
Gabrielyan, A. A., A. I. Adamyan, W. T. Akopyan, S. K. Arsimanyan, A. T. Begun, O. A. Sarkisyan and *G. P. Simonyan:* Tectonic map and map of eruptive rocks of the Armenian SSR. Erewan 1968 (russ.), 73 p.
Garrison, R. E. and *A. G. Fischer:* Deep-water limestones and radiolarites of the Alpine Jurassic. Soc. Ec. Pal. Min. Spec. Publ., Tulsa, 14 (1969), 20–56.
Gass, I. G. and *J. D. Smewing:* Intrusion, extrusion and metamorphism at constructive margins; evidence from the Troodos massif, Cyprus. Nat., London, 242 (1973), 26–29.
Gedikoğlu, A.: Étude géologique de la région de Gölköy (province d'Ordu, Turquie). Thèse Fac. Sc. Univ. Grenoble, Grenoble 1970, 105 p.
Gianelli, G., P. Passerini and *G. Sguazzoni:* Some observations on mafic and ultramafic complexes north of the Bolkardağ (Taurus, Turkey). Boll. Soc. Geol. It., Roma, 91 (1972), 439–488.
Girdler, R. W. and *P. Styles:* Two stage Red Sea floor spreading. Nat., London, 247 (1974), 7–11.
Gjelsvik, T.: Investigations of lead-zinc deposits in northwest Anatolia, Turkey. Bull. MTA 59 (1962), 62–70.
Gökçen, S. L. and *G. Ataman:* Sédimentologie des roches détritiques de la formation de Kesan (Paléogène); un facies à turbidites au Sud-Ouest de la Thrace turque. Sed. Geol., Amsterdam, 9 (1973), 235–260.
Gohlke, P. (Ed.): Aristoteles. Die Lehrschriften – Meteorologie. Paderborn 1955, 192 p.
Grancy, W. S.: Überblick über die bisherigen Aufschlußarbeiten und Ergebnisse im östlichen anatolischen Steinkohlenbecken. MTA Mecm. 4 (4) (1939), 64–88.
Griffith, W. R., I. P. Albers and *Ö. Öner:* Massive sulfide copper deposits of the Ergani Maden area, southeastern Turkey. Ec. Geol., Lancaster, Pa., 67 (1972), 701–716.
Güldalı, N.: Karstmorphologische Studien im Gebiete des Poljesystems von Kestel (Westlicher Taurus, Türkei). Tüb. Geogr. Stud., Tübingen, 40 (1970), 104 p.
Gümüş, A.: Contribution à l'étude géologique du secteur septentrional de Kalabak Köy–Eymirköy (région d'Edremit), Turquie. Publ. MTA 117 (1964), 109 p.
Gümüş, A.: Türkiye metalojenisi. I:2 1/2 mil. Türkiye metalojenetik haritasinin izahi. Publ. MTA 144 (1970), 30 p.
Gümüs, A.: Les minéralisations alpines de la Turquie. Geologija, Ljubljana, 15 (1972), 315–326.
Gümüş, H.: Geologie des mittleren Teiles der Halbinsel Karaburun (Izmir). Scient. Rep. Fac. Sc. Ege Univ., Bornova-Izmir, 100 (1971), 16 p.
Günther, D. and *H. Pichler:* Die obere und untere Bimsstein-Folge auf Santorin. N. Jahrb. Geol. Pal. Monatsh., Stuttgart, 1973, 394–415.
Güvenç, T.: Étude stratigraphique et micropaléontologique du Carbonifère et du Permien des Taurus occidentaux dans l'arrière-pays d'Alanya (Turquie). Thèse Fac. Sc. Univ. Paris, Paris 1965, 292 p.
Güvenç, T.: Description de quelques espèces d'algues calcaires (Gymnocodiacées et Dasycladacées) du Carbonifère et du Permien des Taurus occidentaux (Turquie). Rev. Micropal., Paris, 9 (1966), 94–103.
Güvenç, T.: A propos de la structure de la paroi des Nodosariida et description d'un nouveau genre Alanyana et de quelques nouvelles espèces du Permien de Turquie. Bull. MTA 69 (1967), 34–43.
Güvenç, T.: Un nouveau genre d'algue calcaire du Permien, Embergella n. g. Bull. GST 15 (1972), 21–25.
Güvenç, T.: Gaziantep-Kilis bölgesi stratigrafisi. Doc. Tezi, Fen Fak. Ege Üniv., Bornova-Izmir, 1973, 77 p.
Gugenberger, O.: Beiträge zur Geologie Kleinasiens mit besonderer Berücksichtigung des anatolischen Lias. Sitzber. Ak. Wiss. Math.-nat. Kl. pt. I, Wien, 137 (1928), 259–282.
Gutnic, M., D. Kelter and *O. Monod:* Découverte de nappes de charriage dans le Nord du Taurus occidental (Turquie). C. R. Ac. Sc., Paris, D 266 (1968), 988–991.
Gutnic, M. and *O. Monod:* Une série condensée dans les nappes du Taurus occidental: la série du Boyali Tepe. C. R. Soc. Géol. Fr., Paris, 1970, 166–167.
Gutnic, M. and *Th. Juteau:* Un exemple de coulées volcaniques sous-marines d'âge jurassique dans le Taurus de Pisidie. Sciences Terre, Nancy, 18 (1973), 115–141.

Haas, W.: Das Alt-Paläozoikum von Bithynien (Nordwest-Türkei). N. Jahrb. Pal. Abh., Stuttgart, 131 (1968a), 178–242.
Haas, W.: Trilobiten aus dem Silur und Devon von Bithynien (NW-Türkei). Palaeontogr., Stuttgart, A 130 (1968b), 60–207.

Hafemann, D.: Die Frage des eustatischen Meeresspiegelanstiegs in historischer Zeit. Verh. Deutsch. Geogr.-Tag., Wiesbaden, 32 (1960), 218–231.

Hall, R. and *R. Mason:* A tectonic mélange from the Eastern Taurus mountains, Turkey. Journ. Geol. Soc., London, 128 (1972), 395–397.

Harrison, J. C. and *S. P. Mathur:* Gravity anomalies in Gulf of California. In: *Van Andel, T.* and *G. G. Shor* (Eds.): Marine geology of the Gulf of California. Am. Ass. Petrol. Geol. Mem., Tulsa, 3 (1964), 76–89.

Haude, H.: Das Alt-Paläozoikum – Präkambrium bis Silurium – in der Türkei. Zentralbl. Geol. Pal. pt. I, Stuttgart 1969, 702–719.

Haude, H.: Stratigraphie und Tektonik des südlichen Sultan-Dağ (SW-Anatolien). Zeitschr. Deutsch. Geol. Ges., Hannover, 123 (1972), 411–421.

Hecht, J.: Zur Geologie von Südost-Lesbos (Griechenland). Zeitschr. Deutsch. Geol. Ges., Hannover, 123 (1972), 423–432.

Heritsch, F.: Devonversteinerungen aus dem Antitaurus. N. Jahrb. Geol. Pal. Beil.-Bd., Stuttgart, 59 B (1928), 300–303.

Heritsch, F.: Karbon und Perm in den Südalpen und in Südosteuropa. Geol. Rundsch., Stuttgart, 30 (1939a), 529–588.

Heritsch, F.: Ein Vorkommen von marinem Perm im nördlichen Ala-Dag (Kilikischer Taurus, Türkei). II. Korallen, stratigraphische und paläogeographische Bemerkungen. Sitzber. Ak. Wiss. Math.-nat. Kl. pt. I, Wien, 148 (1939b), 171–194.

Heritsch, F. and *H. R. Von Gaertner:* Devonische Versteinerungen aus Paphlagonien. Sitzber. Ak. Wiss. Math.-nat. Kl. pt. I, Wien, 138 (1929), 189–209.

Hinz, K.: Results of seismic refraction and seismic reflection measurements in the Ionian Sea. Geol. Jahrb., Hannover, E 2 (1974), 33–65.

Höll, R.: Genese und Altersstellung von Vorkommen der Sb-W-Hg-Formation in der Türkei und auf Chios/ Griechenland. Abh. Bayr. Ak. Wiss. Math.-nat. Kl. N. s., München, 127 (1966), 118 p.

Holzer, H. F. and *H. Colin:* Beiträge zur Ophiolithfrage in Anatolien. Jahrb. Geol. Bundesanst., Wien, 100 (1957), 213–237.

Horstink, J.: The Late Cretaceous and Tertiary geological evolution of eastern Turkey. In: *Keskin, C.* and *F. Demirmen* (Eds.): First Petroleum Congress of Turkey, Proc., Ankara 1971, p. 25–41.

Hrouda, B.: Vorderasien I. Mesopotamien, Babylonien, Iran und Anatolien. In: *Hausmann, U.* (Ed.): Handbuch der Altertumswissenschaft, München 1971, 338 p.

Hsü, K. J.: Franciscan mélange as a model for eugeosynclinal sedimentation and underthrusting tectonics. Journ. Geophys. Res., Washington D. C., 76 (1971), 1162–1170.

Hsü, K. J. and *W. B. F. Ryan:* Summary of the evidence for extensional and compressional tectonics in the Mediterranean. In: *Ryan, W. B. F., K. J. Hsü* et al.: Initial reports of the Deep Sea drilling project, Washington, v. 13 (1973), 1011–1019.

Hsü, K. J., M. B. Cita and *W. B. F. Ryan:* The origin of the Mediterranean evaporites. In: *Ryan, W. B. F.* and *K. J. Hsü* et al.: Initial reports of the Deep Sea drilling project, Washington, v. 13 (1973), 1203–1231.

Ilhan, E.: The structural features of Turkey. In: *Campbell, A. S.* (Ed.): Geology and History of Turkey. Tripoli 1971a, p. 159–170.

Ilhan, E.: Earthquakes in Turkey. In: *Campbell, A. S.* (Ed.): Geology and History of Turkey. Tripoli 1971b, p. 431–442.

Irion, G.: Die anatolischen Salzseen, ihr Chemismus und die Entstehung ihrer chemischen Sedimente. Arch. Hydrobiol., Stuttgart, 71 (1973), 517–557.

Irrlitz, W.: Lithostratigraphie und tektonische Entwicklung des Neogens in Nordostanatolien. Beih. Geol. Jahrb., Hannover, 120 (1972), 111 p.

Isgüden, Ö.: Anamur bölgesinin jeolojik etüdü. Tez Fen Fak. Üniv. Istanbul 1968, Istanbul 1971, 85 p.

Ivanov, R. and *K.-O. Kopp:* Das Alttertiär Thrakiens und der Ostrhodope. Geol. et Palaeontolog., Marburg, 3 (1969a), 123–151.

Ivanov, R. and *K.-O. Kopp:* Zur Tektonik des Thrakischen Alttertiär-Beckens. Geotekt. Forsch., Stuttgart, 31 (1969b), 117–132.

Izdar, E.: Geologischer Bau, Magmatismus und Lagerstätten der östlichen Hekimhan-Hasancelebi-Zone (Ostanatolien). Publ. MTA 112 (1963), 72 p.

Izdar, E.: Introduction to geology and metamorphism of the Menderes Massif of western Turkey. In: *Campbell, A. S.* (Ed.): Geology and History of Turkey. Tripoli 1971, p. 495–500.

Jacobshagen, V.: Die Trias der mittleren Ost-Ägäis und ihre paläogeographischen Beziehungen innerhalb der Helleniden. Zeitschr. Deutsch. Geol. Ges., Hannover, 123 (1972), 445–454.

Jacobson, H. S. and *E. Türet:* Geology of the Eymir iron mine, Edremit, Turkey. MTA Rap. 4603 (1970), 28 p. (unpublished).

Janetzko, P.: Geologische Untersuchungen an der Ostflanke des südlichen Amanos-Gebirges zwischen Islahiye und Hassa (Südtürkei). Geotekt. Forsch., Stuttgart, 42 (1972), 3–33.
Jongmanns, W. J.: Beiträge zur Kenntnis der Karbonflora in den östlichen Teilen des Anatolischen Kohlenbeckens. Publ. MTA B 2 (1939), 40 p.
Jongmanns, W. J.: Notes paléobotaniques sur les bassins houillers de l'Anatolie. Med. Geol. Stichting N. S., Maastricht, 9 (1956), 55–89.
Jung, D. and *J. Keller:* Die jungen Vulkanite im Raume zwischen Konya und Kayseri (Zentral-Anatolien). Zeitschr. Deutsch. Geol. Ges., Hannover, 123 (1972), 503–512.
Juteau, Th.: Commentaire de la carte géologique des ophiolites de la région de Kumluca (Taurus lycien, Turquie méridionale). Bull. MTA 70 (1969), 70–91.
Juteau, Th.: Pétrogenèse des ophiolites des nappes d'Antalya (Taurus lycien oriental, Turquie). Sciences Terre, Nancy, 15 (1970), 265–288.
Juteau, Th., H. Lapierre, A. Nicolas, J.-F. Parrot, L.-E. Ricou, G. Rocci and *M. Rollet:* Idées actuelles sur la constitution, l'origine et l'évolution des assemblages ophiolitiques mésogéens. Bull. Soc. Géol. Fr., Paris, VII, 15 (1974), 478–493.

Kahler, F.: Fusuliniden aus T'ien-schan und Tibet. Mit Gedanken zur Geschichte der Fusulinen-Meere im Perm. Rep. Sc. Exped. Northwestern Prov. of China, Stockholm, V, 4 (1974), 148 p.
Kalafatçıoğlu, A.: Geology of the western part of Antalya Bay. Bull. MTA 81 (1973), 31–84.
Kalafatçıoğlu, A. and *H. Uysallı:* Geology of the Beypazarı-Nallıhan-Seben region. Bull. MTA 62 (1964), 1–11.
Kalenić, M.: Cambrian. Carpato-Balkan Geol. Ass., 8. Congr., Belgrade 1967.
Kalkancı, S.: Étude géologique et pétrochimique du Sud de la région de Suşehri. Géochronologie du massif syénitique de Köse Dağ (NE de Sivas, Turquie). Thèse Univ. Grenoble, 1974, 139 p.
Karacabey, N.: Quelques rudistes provenant de la région de Divriği (Turquie orientale). Bull. MTA 78 (1972), 46–54.
Karnik, V.: Seismicity of the European area. Dordrecht, 1971, Pt. 2, 218 p.
Kaya, O.: Karbon bei Istanbul. N. Jahrb. Geol. Pal. Monatsh., Stuttgart 1969, 160–173.
Kaya, O.: The Carboniferous stratigraphy of Istanbul. Bull. GST 14 (1971), 143–199.
Kaya, O.: Tavşanlı yöresi „ofiolit“ sorunun anaçizgileri. Bull. GST 15 (1972a), 26–108.
Kaya, O.: Aufbau und Geschichte einer anatolischen Ophiolith-Zone. Zeitschr. Deutsch. Geol. Ges., Hannover, 123 (1972b), 491–501.
Kaya, O.: Paleozoic of Istanbul. Ege Üniv. Fen Fak. Kitapl. 40, Bornova-Izmir (1973), 143 p.
Kazmin, V. G.: The problem of the „Alpine Mélange“. Geotectonics, Washington, 1971 (2), 73–77.
Keller, J. and *D. Ninkovich:* Tephra-Lagen in der Ägäis. Zeitschr. Deutsch. Geol. Ges., Hannover, 123 (1972), 579–587.
Kemper, E.: Beobachtungen an obereozänen Riffen am Nordrande des Ergene-Beckens (Türkisch-Thrazien). N. Jahrb. Geol. Pal. Abh., Stuttgart, 125 (1966), 540–554.
Keskin, C.: Geology of the Pınarhisar area. Bull. GST 14 (1971a), 31–84.
Keskin, C.: Sedimentary microfacies of the Cretaceous carbonate rock sequence in district V and their importance in stratigraphic correlations. In: *Keskin, C.* and *F. Demirmen* (Eds.): First Petroleum Congress of Turkey, Proc., Ankara 1971b, p. 73–80.
Ketin, I.: Über den geologischen Bau der Şeytandağları und ihrer näheren Umgebung im Nordosten von Tunceli (Ostanatolien). Rev. Ist. 10 (1945), 288–298.
Ketin, I.: Die geologischen Grundzüge der Gegend von Elaziğ (Ostanatolien). Rev. Ist. 12 (1947a), 255–267.
Ketin, I.: Über die Tektonik des Uludağ-Massivs. Bull. GST I (1) (1947b), 59–88.
Ketin, I.: Über die tektonischen Ergebnisse der Geländeaufnahmen des Gebiets Ergani-Eğil in Südost-Anatolien. Rev. Ist. 15 (1950), 153–160.
Ketin, I.: Über die Geologie der Gegend von Bayburt in Nordost-Anatolien. Rev. Ist. 16 (1951), 113–127.
Ketin, I.: Tektonische Untersuchungen auf den Prinzeninseln nahe Istanbul (Türkei). Geol. Rundsch., Stuttgart, 41 (1953), 161–172.
Ketin, I.: Über die Geologie der Gegend von Ovacuma östlich Zonguldak. Rev. Ist. 20 (1955a), 147–154.
Ketin, I.: On the geology of Yozgat region and the tectonic features of the Central Anatolian massif (Kırşehir crystallines). Bull. GST 6 (1) (1955b), 1–40.
Ketin, I.: Über einige meßbare Überschiebungen in Anatolien. Berg- u. Hüttenm. Monatsh., Leoben, 101 (1956), 22–24.
Ketin, I.: Über die Tektonik des Çamlıca-Gebiets bei Istanbul. Bull. GST 7 (1) (1959a), 1–18.
Ketin, I.: The orogenetic evolution of Turkey. Bull. MTA 53 (1959b), 82–88.
Ketin, I.: Über Alter und Art der kristallinen Gesteine und Erzlagerstätten in Zentral-Anatolien. Berg- und Hüttenm. Monatsh., Leoben, 104 (1959c), 163–169.
Ketin, I.: Über die magmatischen Erscheinungen in der Türkei. Bull. GST 7 (2) (1961), 16–33.
Ketin, I.: Tectonic units of Anatolia. Bull. MTA 66 (1966 a), 23–34.

Ketin, I.: Cambrian outcrops in southeastern Turkey and their comparison with the Cambrian of East Iran. Bull. MTA 66 (1966b), 77–89.
Ketin, I.: Relations between general tectonic features and the main earthquake regions of Turkey. Bull. MTA 71 (1968), 63–67.
Ketin, I.: Über die nordanatolische Horizontalverschiebung. Bull. MTA 72 (1969), 1–28.
Khain, V. Y.: The geosynclinal process and the evolution of the tectonosphere. Ak. N. USSR Izv. Geol. ser. 1964, Moskva, 1964 (12), 3–17.
Khain, V. E. and *E. E. Milanovsky:* Structure tectonique du Caucase après les données modernes. Livre Mém. *P. Fallot,* Paris, 2 (1960/63), 663–703.
Kilinc, M.: Étude géologique de la région de l'alpage de Çambaşı (province Ordu, Turquie). Thèse Fac. Sc. Univ. Grenoble 1971, 113 p.
Kines, T.: Geothermometry of the Keban mine area, Eastern Turkey. Bull. MTA 73 (1969), 28–33.
Kıraner, F.: Geology of the eastern region of Lake Van. Bull GST 7 (1) (1959), 30–57.
Klaer, W.: Geomorphologische Untersuchungen in den Randgebirgen des Van-See (Ostanatolien). Zeitschr. Geomorph. N. S., Berlin, 9 (1965), 346–355.
Kleinsorge, H.: Über die allgemeinen geologischen Grundlagen der Standortplanung der neuen Talsperren der Türkei. Geol. Jahrb., Hannover, 78 (1961), 199–236.
Kleinsorge, H. and *R. Vinken:* Beiträge zu den Fragen der Gliederung des Tertiärs in Zentralanatolien. Geol. Jahrb., Hannover, 83 (1965), 209–220.
Klemm, D. D.: Die Eisenerzvorkommen von Divrik (Anatolien) als Beispiel tektonisch angelegter pneumatolytisch-metasomatischer Lagerstättenbildung. N. Jahrb. Min. Abh., Stuttgart, 94 (1960), 591–607.
Knipper, A. L.: Constitution and age of serpentinite mélange in the Lesser Caucasus. Geotectonics, Washington, 1971a (5), 275–281.
Knipper, A. L.: Development of serpentinite mélange in the Lesser Caucasus. Geotectonics, Washington, 1971b (6), 384–389.
Kockel, F., H. Mollat and *H. W. Walter:* Geologie des Serbo-Mazedonischen Massivs und seines mesozoischen Rahmens (Nordgriechenland). Geol. Jahrb., Hannover, 89 (1971), 529–551.
Kogan, I. I., Y. P. Malovitsky, A. P. Milashin, G. V. Osipov and *B. D. Uglov:* New data on the deep-seated structure of the Eastern Mediterranean (based on hydromagnetic survey data). Geotectonics, Washington, 1969 (6), 348–351.
Kopp, K.-O., N. Pavoni and *C. Schindler:* Geologie Thrakiens IV: Das Ergene-Becken. Beih. Geol. Jahrb., Hannover, 76 (1969), 136 p.
Kovenko, V.: Gîtes de magnétite accompagné de tourmaline de la région de Divrik. Publ. MTA B 3 (1939), 100p.
Kovenko, V.: La métallogénie de l'ancien gîte de pyrite cuivreuse de Küre, du gîte nouvellement trouvé d'Asıköy et de la zone côtière (centrale et est) de la Mer Noire. MTA Mecm. 9 (2/32) (1944), 180–211.
Kraeff, A.: Geology and mineral deposits of the Hopa-Murgul region (western part of the province of Artvin, NE-Turkey). Bull. MTA 60 (1963), 45–60.
Kraus, E.: Ein geologisches Gesamtprofil durch die Gebirge Anatoliens Trabzon–Urfa–Harran. MTA Rap. 2521 (1957), 263 p. (unpublished).
Kraus, E.: Die Orogene Ostanatoliens und ihre Schubweiten. Bull. MTA 51 (1958), 1–6.
Kronberg, P.: Photogeologische Daten zur Tektonik im Ostpontischen Gebirge (NE-Türkei). Bull. MTA 74 (1970), 24–33.
Kühn, O.: Korallen des Miozäns von Cilicien. Jahrb. Geol. B.-Anst., Wien, 76 (1926), 65–80.
Kühn, O.: Stratigraphie und Paläogeographie der Rudisten I. Rudistenfaunen und Kreideentwicklung in Anatolien. N. Jahrb. Geol. Pal. Beil.-Bd., Stuttgart, B 70 (1933), 227–250.
Kumbasar, I.: Keban bölgesindeki cevherleşmelerin petrografik ve metalogenetik etüdü. Tez Ist. Tekn. Üniv. Maden Fak., Istanbul 1964, 113 p.
Kurtman, F.: Geologie des Gebietes zwischen Sivas und Divriği sowie Bemerkungen über die Gipsserie. Bull. MTA 56 (1961a), 1–12.
Kurtman, F.: Stratigraphie der Gipsablagerungen im Bereich von Sivas (Zentral-Anatolien). Bull. MTA 56 (1961b), 13–16.
Kurtman, F.: Geologic and tectonic structure of the Sivas–Hafik–Zara and Imranlı region. Bull. MTA 80 (1973a), 1–32.
Kurtman, F. (Ed.): First geothermal energy symposium, Proc., Ankara 1973b, 208 p.
Kurtman, F. and *M. F. Akkus:* Inter-mountain basins in Eastern Anatolia and their oil possibilities. Bull. MTA 77 (1971), 1–9.
Kushan, B.: Stratigraphie und Trilobitenfauna in der Mila-Formation (Mittelkambrium-Tremadoc) im Alborz-Gebirge (N-Iran). Palaeontogr., Stuttgart, A 144 (1973), 113–165.
Kuss, S. E.: Die pleistozänen Säugetierfaunen der ostmediterranen Inseln. Ihr Alter und ihre Herkunft. Ber. Naturf. Ges., Freiburg/Br., 63 (1973), 49–71.

Lahn, E.: La formation gypsifère en Anatolie (Asie Mineure). Bull. Soc. Géol. Fr., Paris, V, 20 (1950), 451–457.

Lahner, L.: Geologische Untersuchungen an der Ostflanke des mittleren Amanos (SE-Türkei). Geotekt. Forsch., Stuttgart, 42 (1972), 64–96.

Laking, Ph. N.: The Black Sea. Its geology, chemistry and biology. A bibliography. Woods Hole Oceanogr. Inst. Contrib., Woods Hole, Mass., 3330 (1974), 368 p.

Lambert, J.: Échinides crétacés de la région d'Héraclée. Ann. Soc. Géol. Belg., Liège, 54 (1931), 3–12.

Lambert, J. and *F. Charles:* Échinides crétacés de la région de Djidde (Anatolie). Bull. Soc. Belg. Géol. Pal. Hydr., Bruxelles, 47 (1937), 377–401.

Lange, S.P.: The subdivision of the Cenozoic in Eastern Central Anatolia. Newsl. Stratigr., Leiden, 1 (3) (1971), 37–40.

Lapierre, H. and *J.-F. Parrot:* Identité géologique des régions de Paphos (Chypre) et du Bair-Bassit (Syrie). C. R. Ac. Sc., Paris, D. 274 (1972), 1999–2002.

Laurentiaux, D.: La faune continentale des marnes de Tchakras, Asie Mineure. Ann. Soc. Géol. Nord, Lille, 66 (1946), 213–234.

Lebling, Cl.: Über eine Reise von Angora nach Ineboli am Schwarzen Meer. In: *Wilser, J. L.* (Ed.): Die Kriegsschauplätze geologisch dargestellt. Berlin, 13 (1925), 104–114.

Lefèvre, R. and *J. Sornay:* Espèces nouvelles d'Inoceramus dans le Taurus lycien (Turquie). Bull. Soc. Géol. Fr., Paris, VII, 8 (1967), 870–876.

Lefèvre, R. and *J. Marcoux:* Schéma structural et esquisse stratigraphique des nappes d'Antalya dans leur segment sud-occidental (Taurus lycien, Turquie). C. R. Ac. Sc., Paris, 271 (1970), 888–891.

Lehnert-Thiel, Kl.: Geologisch-lagerstättenkundliche Untersuchungen an den Zinnobervorkommen Kalecık und dem nordöstlichen Teil der Halbinsel Karaburun (westliche Türkei). Bull. MTA 72 (1969), 43–73.

Le Maître, D.: Observations sur les algues et les foraminifères des calcaires dévoniens. Ann. Soc. Géol. Nord, Lille, 55 (1930), 42–50.

Le Maître, D.: Description des stromatoporoides. Mém. Soc. Géol. Belg., Liège, 1933, 162–165.

Leo, G. W.: Geology and metasomatic iron deposits of the Şamlı region, Balıkesir province, western Turkey. U. S. Geol. Surv. Prof. Pap., Washington, 800 D (1972), 75–87.

Leo, G. W., R. F. Marvin and *H. H. Mehnert:* Geologic framework of the Kuluncak-Sofular area, east-central Turkey, and K-Ar ages of igneous rocks. Bull. Geol. Soc. Am., Boulder, 85 (1974), 1785–1788.

Leuchs, K.: Ladinische und karnische Transgression in Anatolien. N. Jahrb. Min. Geol. Pal. pt. B, Zentralbl., Stuttgart 1939, 303–313.

Leuchs, K.: Der Bauplan von Anatolien. N. Jahrb. Min. Geol. Pal. pt. B, Monatsh., Stuttgart 1943, 33–72.

Lisenbee, A. L.: The Orhaneli ultramafic-gabbro thrust sheet and its surrounding: a progress report. In: *Campbell, A. S.* (Ed.): Geology and History of Turkey. Tripoli 1971, p. 349–368.

Lisenbee, A. L.: Structural setting of the Orhaneli ultramafic massif near Bursa, northwestern Turkey. Thesis Pennsylvania State Univ. 1972, 157 p.

Loboziak, St. and *N. Dil:* Sur l'âge Westphalien inférieur des couches de charbon des mines de Çaydamar (Turquie). Rev. Paléobot. Pal., Amsterdam, 15 (1973), 287–299.

Löffler, E.: Untersuchungen zum eiszeitlichen und rezenten klimagenetischen Formenschatz in den Gebirgen Nordostanatoliens. Heidelb. Geogr. Arb., Heidelberg, 27 (1970), 126 p.

Lort, J. M. and *F. Gray:* Cyprus, seismic studies at sea. Nat., London, 248 (1974), 745–747.

Lort, J. M., W. Q. Limond and *F. Gray:* Preliminary seismic studies in the eastern Mediterranean. Earth Plan. Sc. L., Amsterdam, 21 (1974), 355–366.

Louis, H.: Eiszeitliche Seen in Anatolien. Zeitschr. Ges. Erdk., Berlin 1938, 267–285.

Louis, H.: Die Spuren eiszeitlicher Vergletscherung in Anatolien. Geol. Rundsch., Stuttgart, 34 (1944), 447–481.

Lüttig, G. and *P. Steffens:* Paleogeographic atlas of Turkey from the Oligocene to the Pleistocene. Geol. Jahrb. B, Hannover (in preparation).

Lys, M.: Les calcaires à fusulines des environs de Bergama (Turquie): Zeytindağ et Kınık. Notes Mém. Moyen-Orient, Paris, 12 (1971/72), 167–171.

Magné, J. and *A. Poisson:* Présence de niveaux oligocènes dans les formations sommitales du massif des Bey Dağları près de Korkuteli et de Bucak (Autochtone du Taurus lycien, Turquie). C. R. Ac. Sc., Paris, D 278 (1974), 205–208.

Makris, J.: Some geophysical aspects of the evolution of the Hellenides. Bull. Geol. Soc. Greece, Athens, 10 (1973), 206–213.

Maley, T. S. and *G. L. Johnson:* Morphology and structure of the Aegean Sea. Deep-Sea Res., Oxford–London–New York–Paris, 18 (1971), 109–122.

Marcoux, J.: Alpine type Triassic of the upper Antalya nappe (western Taurides, Turkey). In: *Zapfe, H.* (Ed.): Die Stratigraphie der alpin-mediterranen Trias, Wien 1974, 145–146.

Marcoux, J. and *A. Poisson:* Une nouvelle unité structurale majeure dans les nappes d'Antalya: la nappe inférieure et ses séries mésozoiques radiolaritiques (Taurides occidentales, Turquie). C. R. Ac. Sc., Paris, D 275 (1972), 655–658.
Martin, C.: Étude stratigraphique et tectonique d'une partie du Taurus au nord d'Akseki (Turquie méridionale). Bull. MTA 72 (1969), 110–129.
Maucher, A., H.-H. Schultze-Westrum and *H. Zankl:* Geologisch-lagerstättenkundliche Untersuchungen im Ostpontischen Gebirge. Abh. Bayr. Ak. Wiss. Math.-nat. Kl. N. S., München, 109 (1962), 97 p.
Maxwell, J. C.: Anatomy of an orogen. Bull. Geol. Soc. Am., Boulder, 85 (1974), 1195–1204.
McKenzie, D.: Active tectonics of the Mediterranean region. Geophys. Journ., Oxford, 30 (1972), 109–185.
Mercier, J.: Sur l'âge des calcaires noirs de Çarçal Dağ (Anatolie sud-orientale). C. R. Soc. Géol. Fr., Paris, 1951, 257–258.
Meric, E.: Sur quelques Loftusiidae et Orbitoididae de la Turquie. Rev. Ist. 32 (1967), 1–58.
Messerli, B.: Die eiszeitliche und die gegenwärtige Vergletscherung im Mittelmeerraum. Geogr. Helv., Bern, 22 (1967), 105–228.
Metz, K.: Ein Vorkommen von marinem Perm im nördlichen Ala Dagh (Kilikischer Taurus, Türkei). I. Allgemeines, Brachiopoden und Bryozoen. Sitzber. Ak. Wiss. Math.-nat. Kl. pt. I, Wien, 148 (1939a), 141–152.
Metz, K.: Beiträge zur Geologie des Kilikischen Taurus im Gebiet des Ala Dagh. Sitzber. Ak. Wiss. Math.-nat. Kl. pt. I, Wien, 148 (1939b), 287–340.
Metz, K.: Zur Paläontologie des anatolischen Paläozoikums I. Neufunde im Paläozoikum SW-Anatoliens. N. Jahrb. Geol. Pal. Abh., Stuttgart, 101 (1955), 257–266.
Meulenkamp, J. E.: The Neogene in the southern Aegean area. In: *Strid, A.* (Ed.): Evolution in the Aegean, Opera Botan., Lund, 30 (1971), 5–12.
Milanovsky, E. E.: Problems of origin of Black Sea depression and its position in the structure of the Alpine belt. Int. Geol. Rev., Washington, 9 (1967), 1237–1249.
Milsom, J.: Papuan ultramafic belt: gravity anomalies and the emplacement of ophiolites. Bull. Geol. Soc. Am., Boulder, 84 (1973), 2243–2258.
Molnar, P.: Fault plane solutions of earthquakes and direction of motion in the Gulf of California and in the Rivera fracture zone. Bull. Geol. Soc. Am., Boulder, 84 (1973), 1651–1658.
Monod, O.: Présence d'une faune ordovicienne dans les schistes de Seydisehir à la base des calcaires du Taurus occidental. Bull. MTA 69 (1967), 79-89.
Monod, O., J. Marcoux, A. Poisson and *J.-F. Dumont:* Le domaine d'Antalya, témoin de la fracturation de la plateforme africaine au cours du Trias. Bull. Soc. Géol. Fr., Paris, VII, 16 (1974), 116–127.
Moore, D. G.: Plate-edge deformation and crustal growth, Gulf of California structural province. Bull. Geol. Soc. Am., Boulder, 84 (1973), 1883–1906.
Moores, E. M.: Geotectonic significance of ultramafic rocks. Earth-Sc. Rev., Amsterdam, 9 (1973), 241–258.
MTA: Geological map of Turkey 1:500 000, 18 sheets w. explanations. 1961/64.
Müller-Karpe, H.: Handbuch der Vorgeschichte v. I, Altsteinzeit. München 1966, 389 p.
Muratov, M. V.: Structure and evolution of the folded basement of the Mediterranean belt of Europe and western Asia. Geotectonics, Washington 1969, 71–79.
Muratov, M. V. and *Y. P. Neprochnov:* Structure of the Black Sea depression and its origin. Byull. Mosk. obsh. isp. prir., otd. geol., Moskva, 77 (5) (1967), 40–59.
Mutti, E., G. Orombelli and *R. Pozzi:* Geological studies on the Dodekanes islands (Aegean Sea) IX. Geological map of the island of Rhodes (Greece). Ann. Géol. P. Hellén., Athènes, 22 (1970), 77–226.

Nakoman, E.: Contribution à l'étude palynologique des formations tertiaires du bassin de Thrace. Ann. Soc. Géol. Nord, Lille, 86 (1966a), 65–107.
Nakoman, E.: Analyse sporopollinique des lignites éocènes de Sorgun (Yozgat, Turquie). Bull. MTA 67 (1966b), 68–88.
Nakoman, E.: Quelques formes nouvelles provenant de la microflore tertiaire du sud-ouest de l'Anatolie. Bull. MTA 68 (1967), 27–38.
Nasr, S. N.: Science and civilisation in Islam. Cambridge, Mass. 1968, 384 p.
Nebert, K.: Ein Beitrag zum jüngsten geologisch-tektonischen Werdegang Inneranatoliens. Nachweis der walachischen Orogenphase im Vilayet Ankara (bei Kayı-Bucuk). Bull. MTA 50 (1958), 15–26.
Nebert, K.: Vergleichende Stratigraphie und Tektonik der lignitführenden Neogengebiete westlich und nördlich von Tavşanlı. Bull. MTA 54 (1960), 8–37.
Nebert, K.: Der geologische Bau der Einzugsgebiete Kelkit Çay und Kızılırmak (NE-Anatolien). Bull. MTA 57 (1961a), 1–51.
Nebert, K.: Zur Kenntnis des neogenen Vulkanismus im Raume westlich Gördes (Westanatolien). Bull. MTA 57 (1961b), 52–56.

Nebert, K.: Ein Anthrazitvorkommen im Liasflysch bei Şiran (Vilayet Gümüshane). Bull. MTA 60 (1963), 7–13.
Nebert, K.: Nordbewegungen im südwestlichen Taurus (bei Akseki). Bull. MTA 62 (1964a), 12–41.
Nebert, K.: Zur Geologie des Kelkit Çay-Oberlaufs südwestlich von Şiran (Nordostanatolien). Bull. MTA 62 (1964b), 42–59.
Needham, H. D., X. Le Pichon, M. Melguen, G. Pautot, V. Renard, F. Avedik and *D. Carre:* North Aegean Sea trough. Abstract, 23. Congr. Int. Expl. Mer Méditerr., Athènes 1972.
Neprochnov, Yu. P., A. F. Neprochnova and *Ye. G. Mirlin:* Deep structure of the Black Sea basin. In: *Degens, E. T.* and *D. A. Ross* (Eds.): The Black Sea, geology, chemistry and biology. Am. Ass. Petr. Geol. Mem., Tulsa, 20 (1974), 35–49.
Niehoff, W.: Bolu bölgesi 1:100 000 ölcekli jeoloji haritası. MTA Rap. 1960 (unpublished).
Niklewski, J. and *W. Van Zeist:* A late Quaternary pollen diagram from northwestern Syria. Acta Bot. Neerl., Wageningen, 19 (1970), 737–754.
Ninkovich, D. and *J. D. Hays:* Mediterranean island arcs and origin of high potash volcanoes. Earth Plan. Sc. L., Amsterdam, 16 (1972), 331–345.
Nöth, L.: Oberkreidefossilien aus Paphlagonien (Kleinasien). N. Jahrb. Min. Geol. Pal. Abh., Stuttgart, B 65 (1931), 321–362.
Norman, T.: Stratigraphy of Upper Cretaceous–Lower Tertiary strata of Yahşıhan area, east of Ankara. Bull. GST 15 (1972), 180–276.
Norman, T.: Late Cretaceous/Early Tertiary development of Ankara Yahşıhan region. Bull. GST 16 (1) (1973a), 41–66.
Norman, T.: Post-Eocene tectonic development of Ankara Yahşıhan region. Bull. GST 16 (1) (1973b), 67–81.
Nowack, E.: Die wichtigsten Ergebnisse meiner anatolischen Reisen. Zeitschr. Deutsch. Ges., Berlin, 80 (1928), B 304–B 312.
Nowack, E.: Längs Anatoliens Nordküste. Zeitschr. Ges. Erdk., Berlin 1929, 1–12.
Nowroozi, A. A.: Focal mechanisms of earthquakes in Persia, Turkey, West Pakistan and Afghanistan, and plate tectonics of the Middle East. Bull. Seism. Soc. Am., Berkeley, 62 (1972), 823–850.

Önay, T. S.: Über die Smirgelgesteine Südwest-Anatoliens. Schweiz. Min.-Petr. Mitt., Zürich, 29 (1949), 357–491.
Öngür, T.: Geological development of the country near Isparta in western Taurus mountains. Abstract, Congr. Earth Sc., Ankara 1973.
Özelçi, H. F.: Gravity anomalies of the Eastern Mediterranean. Bull. MTA 80 (1973), 54–92.
Özgül, N.: The importance of block movements in structural evolution of the northern part of central Taurus. Bull. GST 14 (1971), 85–101.
Özgül, N., S. Metin and *W. T. Dean:* Lower Paleozoic stratigraphy and faunas of the Eastern Taurus mountains. Bull. MTA 79 (1972), 9–16.
Özgül, N., S. Metin and *W. T. Dean:* Doğu Toroslar da Tufanbeyli ilçesi (Adana) dolayının Alt Paleozoik stratigrafisi ve fauna. MTA Derg. 79 (1972), 9–16.
Özgül, N., S. Metin, E. Göger, I. Bingöl, O. Baydar and *B. Erdoğan:* Cambrian–Tertiary rocks of the Tufanbeyli region, eastern Taurus, Turkey. Bull. GST 16 (1) (1973), 82–100.
Özgül, N. and *I. Gedik:* New data on the stratigraphy and the conodont faunas of Çaltepe limestone and Seydeşehir formation, Lower Paleozoic of Central Taurus range. Bull. GST 16 (2) (1973), 39–52, Ankara.
Özkaya, I.: Stratigraphy of Sason and Baykan areas, SE-Turkey. Bull. GST 17 (1) (1974), 51–72.
Özpeker, I.: Batı Anadolu borat yataklarının mukayesi jenetik etüdü. Tez Ist. Tekn. Üniv. Maden Fak., Istanbul 1969, 116 p.
Özsayar, T.: Paläontologie und Geologie des Gebietes östlich Trabzon (Anatolien). Gieß. Geol. Schrift., Gießen, 1 (1971), 138 p.
Öztemür, V.: Note on some foraminiferal species encountered in the well samples of southeastern Turkey. Bull. MTA 52 (1959), 59–66.
Öztunali, Ö.: Uludag (kuzeybatı Anadolu) ve Egrigöz (batı Anadolu) masiflerinin petrolojileri ve geokronologileri. Ist. Mon. 23 (1973), 115 p.
Oğuz, M.: Çaldag'da (Manisa) jeolojik bir araştırma. MTA Derg. 68 (1967), 102–105.
Okay, A. C.: Geologische und petrographische Untersuchung des Gebietes zwischen Alemdağ, Karlıdağ und Kayışdağ in Kocaeli (Bithynien, Türkei). Rev. Ist. 12 (1947), 269–287.
Okay, A. C.: Geologische Untersuchungen des Gebiets zwischen Kayseri, Niğde und Tuzgölü. Rev. Ist. 22 (1957), 53–70.
Oppenheim, P.: Das Neogen in Kleinasien. Zeitschr. Deutsch. Geol. Ges., Berlin, 70 (1919), 1–210.

Orombelli, G., G. P. Lozej and *L. A. Rossi:* Preliminary notes on the geology of the Datca peninsula (SW-Turkey). Rend. Acc. Naz. Linc. Cl. Sc. Fis. Mat. Nat., Roma, VIII, 42 (1967), 830–841.
Oswald, F.: Armenien. In: *Steinmann, G.* and *O. Wilckens* (Eds.): Handb. Reg. Geol., Heidelberg, 10 (1912), 40 p. Reprint Meisenheim 1968.
Ott, E.: Mitteltriadische Riffe der Nördlichen Kalkalpen und altersgleiche Bildungen auf Karaburun und Chios (Ägäis). Mitt. Ges. Geol. Bergbaustud., Innsbruck, 21 (1972), 251–276.
Ovalıoğlu, R.: Die Chromerzlagerstätten des Pozantı-Reviers und ihre ophiolithischen Muttergesteine. Publ. MTA 114 (1963), 85 p.
Ozansoy, F.: Faune des mammifères du Tertiaire de Turquie et leurs révisions stratigraphiques. Bull. MTA 49 (1957), 29–48.
Ozansoy, F.: Resultats essentiels de l'étude de la succession faunistique de la région d'Ankara (Turquie). Bull. MTA 56 (1961), 50–60.
Ozansoy, F.: Les Anthracothériens de l'Oligocène inférieur de la Thrace orientale (Turquie). Bull. MTA 58 (1962), 85–96.
Ozansoy, F.: Étude des gisements continentaux et des mammifères du Cénozoique de Turquie. Mém. Soc. Géol. Fr. N. S., Paris, 102 (1965), 92 p.
Ozansoy, F.: Pleistocene fossil human footprints in Turkey. Bull. MTA 72 (1969), 146–150.

Paeckelmann, W.: Beiträge zur Kenntnis des Devons am Bosporus, im besonderen in Bithynien. Abh. Preuss. Geol. Landesanst. N. S., Berlin, 98 (1925), 152 p.
Paeckelmann, W.: Neue Beiträge zur Kenntnis der Geologie, Paläontologie und Petrographie der Umgegend von Konstantinopel. II. Geologie Thraziens, Bithyniens und der Prinzeninseln. Abh. Preuss. Geol. Landesanst. N. S., Berlin 186 (1938), 202 p.
Pamir, H. N.: Turquie. Lex. Stratigr. Int., Paris 1960, v. 3, pt. 9c, 96 p.
Pamir, H. N. and *F. Baykal:* Contribution à l'étude géologique de la région de Bingöl. Rev. Ist. 8 (1943), 311–318.
Pamir, H. N. and *F. Baykal:* Le massif de Stranca. Bull. GST 1 (1) (1947), 7–43.
Papazachos, B. C.: Distribution of seismic foci in the Mediterranean and surrounding area and its tectonic implication. Geophys. Journ., Oxford, 33 (1973), 421–430.
Paréjas, E.: Le substratum ancien du Taurus occidental au sud d'Afyon Karahissar (Turquie). C. R. Soc. Phys. Hist. Nat., Genève, 60 (1943), 110–114.
Parrot, J.-F.: Pétrologie de la coupe du Djebel Moussa, massif basique-ultrabasique du Kizil Dağ (Hatay, Turquie). Sciences Terre, Nancy, 18 (1973), 143–172.
Pasquaré, G.: Geology of the Cenozoic volcanic area of Central Anatolia. Mem. Acc. Naz. Linc., Cl. Sc. Fis. Math. Nat., Roma, VIII, 9 (1968), 55–204.
Pasquaré, G.: Cenozoic volcanics of the Erzurum area. Geol. Rundsch., Stuttgart, 60 (1971), 900–911.
Passerini, P. and *G. Sguazzoni:* Ricerce sulle ofioliti delle catene alpine. 2. Giacitura delle ofioliti nella zone a sud-ovest de Konya (Anatolia meridionale). Bull. Soc. Geol. It., Roma, 85 (1966), 509–523.
Patijn, R. J. H.: Zonguldak–Kozlu area of the North Anatolian coalfield. Maden, Zonguldak, 20/21 (1953/54), 1–20.
Patijn, R. J. H.: The geology of the Kandilli–Armutcuk coalfield. Maden, Zonguldak, 20/21 (1953/54), 21–28.
Pavoni, N.: Die nordanatolische Horizontalverschiebung. Geol. Rundsch., Stuttgart, 51 (1961), 122–139.
Payne, B. R. and *T. Dincer:* Isotope survey of the karst region of Southern Turkey. 6. Int. Conf. Radiocarbon Tritium dating, Proc. p. 671–686. Pullman, Wash. 1965.
Pejatović, S.: Metallogenetic zones in the Eastern Black Sea – Minor Caucasus region and distinguishing features of their metallogeny. Bull. MTA 77 (1971), 10–22.
Penck, W.: Die tektonischen Grundzüge Westkleinasiens. Stuttgart 1918, 120 p.
Penck, W.: Grundzüge der Geologie des Bosporus. Veröff. Inst. Meeresk. N. S., Berlin, 4 (1919), 71 p.
Penecke, K. A.: Das Sammelergebnis Dr. Franz Schaffers aus dem Oberdevon von Hadschin im Antitaurus. Jahrb. Geol. Reichsanst., Wien, 53 (1903), 141–152.
Perrin, Y.: Étude préliminaire sur les associations minérales des ophiolites des nappes d'Antalya (Taurus lycien et oriental, Turquie). Thèse Fac. Sc. Univ. Nancy, Nancy 1970, 109 p.
Petrascheck, W. E.: Beziehungen zwischen der Anatolischen und der Südosteuropäischen Metallprovinz. Bull. MTA 46/47 (1954/55), 64–74.
Petrascheck, W. E.: Kohlengeologische Probleme im Revier von Zonguldak (Türkei). Berg- u. Hüttenm. Monatsh., Wien, 100 (1955), 70–73.
Petrascheck, W.E.: Die alpin-mediterrane Metallogenese. Geol. Rundsch., Stuttgart, 53 (1964), 376–389.
Petrascheck. W. E.: Die Blei-Zinklagerstätten in Kalken des westlichen Taurus. Bull. MTA 68 (1967), 39–50.
Pfannenstiel, M.: Die altsteinzeitlichen Kulturen Anatoliens. Istanb. Forsch., Berlin, 15 (1941), 50 p.
Pfannenstiel, M.: Die diluvialen Entwicklungsstadien und die Urgeschichte von Dardanellen, Marmarameer und Bosporus. Geol. Rundsch., Stuttgart, 34 (1944), 341–434.

Philippson, A.: Reisen und Forschungen im westlichen Kleinasien. Peterm. Geogr. Mitt. Erg.-H., Gotha, 167, 172, 177, 180, 183 (1910/15), 598 p., 6 sheets 1:300 000.
Philippson, A.: Das Vulkangebiet von Kula in Lydien, die Katakekaumene der Alten. Peterm. Geogr. Mitt., Gotha, 59 (2) (1913), 237–241.
Philippson, A.: Zusammenhang der griechischen und kleinasiatischen Faltengebirge. Peterm. Geogr. Mitt., Gotha, 60 (2) (1914), 71–75.
Philippson, A.: Kleinasien. In: *Steinmann, G.* and *O. Wilckens* (Eds.): Handb. Reg. Geol., Heidelberg, 22 (1918), 183 p., Reprint London–New York 1968.
Philippson, A.: Zur morphologischen Karte des westlichen Kleinasien. Peterm. Geogr. Mitt., Gotha, 66 (1920), 197–202.
Phillips, R. P.: Seismic refraction studies in Gulf of California. In: *Van Andel, T.* and *G. G. Shor* (Eds.): Marine geology of the Gulf of California. Am. Ass. Petrol. Geol. Mem., Tulsa, 3 (1964), 90–121.
Pia, J.: Über eine mittelliassische Cephalopodenfauna aus dem nordöstlichen Kleinasien. Ann. Naturhist. Hofmus., Wien, 27 (1913), 335–388.
Pınar, N.: Les lignes sismiques du bassin égéen de l'Anatolie et les sources thermales. Rev. Fac. Sc. Univ. Ist., Istanbul, A 14 (1949), 20–43.
Pınar, N.: Vue d'ensemble sur les faunes échinologiques de Turquie. C. R. 19. Congr. Géol. Int., Alger, pt. 19 (1954), 109–113.
Pınar, N.: Sur quelques échinides du Crétacé supérieur de la région de Kandıra (Kocaeli, Turquie). Rev. Ist. 21 (1956), 183–189.
Pişkin, Ö.: Étude minéralogique et pétrographique de la région située à l'est de Çelikhan (Taurus oriental, Turquie). Thèse Fac. Sc. Univ. Genève, Genève 1972, 152 p.
Poisson, A.: Présence d'un Trias supérieur de faciès récifal dans le Taurus lycien en NW d'Antalya (Turquie). C. R. Ac. Sc., Paris, 264 (1967), 2443–2446.
Poisson, A.: L'unité inférieure (unité I) du Domuz Dağ (Taurus lycien, Turquie), série sédimentaire avec intercalation de coulées sous-marines en coussins. Bull. MTA 70 (1968), 100–105.
Poisson, A.: Présence de Jurassique et de Crétacé inférieur à faciès de type plate-forme dans l'autochtone lycien près d'Antalya (massif des Bey Dağları, Turquie). C. R. Ac. Sc., Paris, D 278 (1974), 835–838.
Pollak, A.: Über einige geologische Beobachtungen im zentral-anatolischen Massiv. Notizbl. Hess. Landesamt Bodenfschg., Wiesbaden, 87 (1958), 239–245.
Pollak, A.: Die nordanatolische Erzprovinz. Bull. MTA 70 (1968), 92–99.

Rabinowitz, P. D. and *W. B. F. Ryan:* Gravity anomalies and crustal shortening in the eastern Mediterranean. Tectonophys., Amsterdam, 10 (1970), 585–608.
Radelli, L.: La nappe de Balya. La zone des plis égéens et l'extension de la zone du Vardar en Turquie occidentale. Géol. Alpine, Grenoble, 46 (1970), 169–175.
Radelli, L.: Sur la tectonique de la châine anatolienne de Bitlis. In: *Choubert, G.* and *A. Faure-Muret* (Eds.): Tectonique de l'Afrique, Paris 1971, p. 131–139.
Ralli, G.: Le bassin houillier d'Héraclée. Ann. Soc. Géol. Belg., Liège, 23 (1895/96), 151–267.
Ralli, G.: Le bassin houillier d'Héraclée et la flore du Culm et du Houillier Moyen. Paris–Liège 1933, 166 p.
Ramberg, I. B. and *S. B. Smithson:* Gravity interpretation of the southern Oslo Graben and adjacent Precambrian rocks, Norway. Tectonophys., Amsterdam, 11 (1971), 419–431.
Rechinger, K. H.: Grundzüge der Pflanzenverbreitung in der Ägäis. Vegetatio, Den Haag, 2 (1950), 55–119, 239–308, 365–386.
Rezanov, I. A. and *S. S. Chamo:* Reasons for the absence of a „granite" layer in the basins of the South Caspian and the Black Sea. Canad. Journ. Earth Sc., Ottawa, 6 (1969), 671–678.
Richter, M.: Über Zusammenhänge der Gebirge im östlichen Mittelmeer. N. Jahrb. Geol. Pal. Monatsh., Stuttgart 1966, 73–87.
Ricou, L.-E.: Relations entre stades paléogéographiques et phases tectoniques successifs sur l'exemple des Zagrides. Bull. Soc. Géol. Fr., Paris, VII, 15 (1973), 612–623.
Rigassi, D.: Petroleum geology of Turkey. In: *Campbell, A. S.* (Ed.): Geology and History of Turkey. Tripoli 1971, p. 453–482.
Rigo De Righi, M. and *A. Cortesini:* Gravitiy tectonics in foothill structure belt of Southeast Turkey. Bull. Am. Ass. Petr. Geol., Tulsa, 48 (1964), 1911–1937.
Roloff, A.: Die jungkretazisch-tertiäre Entwicklung am W-Rande des Amanosgebirges (Südtürkei). Geotekt. Forsch., Stuttgart, 42 (1972), 97–129.
Roman, J.: Échinides (Clypeaster, Scutella, Schizaster) de l'Helvétien du bassin de Karaman (Turquie). Bull. MTA 55 (1960), 61–96.
Ross, D. A., E. Uchupi and *C. O. Bowin:* Shallow structure of Black Sea. In: *Degens, E. T.* and *D. A. Ross* (Eds.): The Black Sea, geology, chemistry and biology. Am. Ass. Petr. Geol. Mem., Tulsa, 20 (1974), 11–34.

Ross, D. A. and *E. T. Degens:* Recent sediments of Black Sea. In: *Degens, E. T.* and *D. A. Ross* (Eds.): The Black Sea, geology, chemistry and biology. Am. Ass. Petr. Geol. Mem., Tulsa, 20 (1974), 183–199.
Rubinstein, M.: Regionale und lokale Verjüngung des Argon-Alters am Beispiel des Kaukasus. Ecl. Geol. Helv., Basel, 63 (1970), 281–289.
Rückert-Ülkümen, N.: Fossile Fische aus dem Sarmat von Pinarhisar (Türkisch-Thrakien). Senckenberg. Leth., Frankfurt, 46a (1965), 315–361.
Rusnak, G. A., R. L. Fisher and *F. P. Shepard:* Bathymetry and faults of Gulf of California. In: *Van Andel, T.* and *G. G. Shor* (Eds.): Marine geology of the Gulf of California. Am. Ass. Petrol. Geol. Mem., Tulsa, 3 (1964), 59–75.
Russell, R. J.: Alluvial morphology of Anatolian rivers. Ann. Ass. Am. Geogr., Lawrence, Kans., 44 (1954), 363–391.
Ryan, C. W.: A guide to the known minerals of Turkey. MTA, Ankara 1960, 2nd ed., 196 p.
Ryan, W. B. F., D. J. Stanley, J. B. Hersey, D. A. Fahlquist and *Th. D. Allan:* The tectonics of the Mediterranean Sea. In: *Maxwell, A. E.* (Ed.): The Sea. New York–London–Sidney–Toronto 1970, v. 4, pt. 2, 387–492.
Ryan, W. B. F., K. J. Hsü et al. (Eds.): Initial reports of the Deep Sea drilling project. Washington 1973a, v. 13, 2 parts, 1447 p.
Ryan, W. B. F. et al.: Hellenic Trough, site 127 and 128; Strabo Trench and Mountains, site 129; Mediterranean Ridge, Levantine Sea, site 130. In: *Ryan, W. B. F., K. J. Hsü* et at. (Eds.): Intitial reports of the Deep Sea drilling project. Washington 1973b, v. 13, 243–382.

Salomon-Calvi, W.: Die Entstehung der anatolischen „Ova“. Arb. Y. Ziraat Enst., Ankara, 30 (1936), 11 p.
Sanver, M.: A paleomagnetic study of Quaternary volcanic rocks from Turkey. Phys. Earth Planet. Int., Amsterdam, 1 (1968), 403–421.
Savasçin, Y.: Beiträge zur Frage der Genese westanatolischer „Andesite“ und „Basalte“. Bull. GST 17 (1) (1974), 87–173.
Sayar, C.: Boğazıcı arazisinde Ordovisien Conularia'ları. Bull. GST 12 (1969), 140–159.
Schäfer, J. and *H. Schläger:* Zur Seeseite von Kyme in der Aeolis. Archäolog. Anz., Berlin 1962, 41–58.
Schiettecatte, J. P.: Geology of the Misis mountains. In: *Campbell, A.S.* (Ed.): Geology and History of Turkey. Tripoli 1971, p. 305–312.
Schmidt, G. C.: A review of Permian and Mesozoic formations exposed near the Turkey/Iraq border at Harbol. Bull. MTA 62 (1964), 103–119.
Schmidt, W. J.: Geologie und Erzführung der Chromitkonzession Başören (Anatolien). Sitzber. Österr. Ak. Wiss. Math.-nat. Kl. pt. I, Wien, 163 (1954), 621–644.
Scholten, R.: Role of the Bosporus in Black Sea chemistry and sedimentation. In: *Degens, E. T.* and *D. A. Ross* (Eds.): The Black Sea, geology, chemistry and biology. Tulsa 1974, 115–126.
Schuiling, R. D.: Über eine prä-herzynische Faltungsphase im Kaz Dağ-Kristallin. Bull. MTA 53 (1959), 89–93.
Schuiling, R. D.: On petrology, age and structure of the Menderes migmatite complex (SW-Turkey). Bull. MTA 58 (1962), 71–84.
Schuiling, R. D.: Active role of continents in tectonic evolution, geothermal models. In: *De Jong, K. A.* and *R. Scholten* (Eds.): Gravity and tectonics. New York–Sidney–London–Toronto 1973, 35–47.
Schwan, W.: Ergebnisse neuerer geologischer Forschungen im Amanosgebirge (Süd-Türkei). Geotekt. Forsch., Stuttgart, 42 (1972), 130–160.
Scotford, D. M.: Metasomatic augen gneiss in greenschist facies, western Turkey. Bull. Geol. Soc. Am., Boulder, 80 (1969), 1079–1094.
Sellier De Civrieux, J. M. and *T. F. J. Dessauvagie:* Reclassification de quelques Nodosariidae, particulièrement du Permien au Lias. Publ. MTA 124 (1965), 179 p.
Serbin, A.: Bemerkungen Strabos über den Vulkanismus und Beschreibung der den Griechen bekannten vulkanischen Gebiete. Diss. Phil. Fak. Univ. Erlangen, Berlin 1893, 96 p.
Sestini, G.: The relations between flysch and serpentinites in north-central Turkey. In: *Campbell, A. S.* (Ed.): Geology and History of Turkey. Tripoli 1971, p. 369–383.
Sestini, G. and *P. Canuti:* Flysch facies in the Pontic Mountains of Turkey. Boll. Soc. Geol. It., Roma, 87 (1968), 317–332.
Seyman, I. and *A. Aydin:* The Bingöl earthquake fault and its relation to the North Anatolian fault zone. Bull. MTA 79 (1972), 1–8.
Sickenberg, O.: Die Gliederung des höheren Jungtertiärs und Altquartärs in der Türkei nach Vertebraten und ihre Bedeutung für die internationale Neogen-Stratigraphie. Geol. Jahrb., Hannover, B 15 (1975a), 167 p.
Sickenberg, O.: Über das Villafranchium in der Türkei. Mém. B. R. G. M., Paris, 78 (1) (1975b), 241–245.
Sirel, E.: Systematic study of new species of the genera Fabularia and Kathina from Paleocene. Bull. GST 15 (1972), 277–294.

Sirel, E.: Description of a new Cuvillierina species from the Maestrichtian of Cide (Northern Turkey). Bull. GST 16 (2) (1973), 69–76.
Skinner, J. K.: Permian foraminifera from Turkey. Univ. Kansas Pal. Contrib., Lawrence, Kans., 36 (1969), 14 p.
Smith, A. G.: Alpine deformation and the oceanic areas of the Tethys, Mediterranean and Atlantic. Bull. Geol. Soc. Am., Boulder, 82 (1971), 2039–2070.
Sönmez-Gökçen, N.: Étude paléontologique (Ostracodes) et stratigraphique de niveaux du Paléogène du Sud-Est de la Thrace. Publ. MTA 147 (1973a), 118 p.
Sönmez-Gökçen, N.: Géologie du bassin d'Ergene et des chaînes de la bordure de la mer de Marmara. Publ. MTA 148 (1973b), 17 p.
Somin, M. L. and *A. A. Belov:* On the tectonic history of the southern slope of the Greater Caucasus. Geotectonics, Washington, 1967 (1), 37–39.
Staesche, U.: Die Geologie des Neogen-Beckens von Elbistan (Türkei) und seiner Umrandung. Geol. Jahrb., Hannover, B 4 (1972), 3–52.
Stanley, D. J. (Ed.): The Mediterranean Sea. A natural sedimentation laboratory. Stroudsburg, Penns. 1972, 765 p.
Stanley, D. J.: Basin plains in the eastern Mediterranean, significance in interpreting ancient marine deposits I. Basin depth and configuration. Mar. Geol., Amsterdam, 15 (1973), 295–307.
Stark, H.: Das frühgeschichtliche Alter des Amiksees bei Antakya (Kleinasien). N. Jahrb. Geol. Pal. Monatsh., Stuttgart 1956, 244–248.
Stchépinsky, V.: Faune Miocène du vilayet de Sivas (Turquie). Publ. MTA C 1 (1939), 63 p.
Stchépinsky, V.: Découverte du Paléocène en Turquie. MTA Mecm. 6 (2/23) (1941), 143–158.
Stchépinsky, V.: Contribution à l'étude de la faune crétacée de la Turquie. Publ. MTA B 7 (1942), 68 p.
Stchépinsky, V.: Géologie de la région de Maras–Gaziantep. MTA Mecm. 8 (1/29) (1943), 110–125.
Stchépinsky, V.: Géologie et ressources minérales de la région de Malatya (Turquie). MTA Mecm. 9 (1/31) (1944), 79–104.
Stchépinsky, V.: Fossiles charactéristiques de Turquie. Publ. MTA D 1 (1946), 151 p.
Stchépinsky, V.: Paléobiogéographie de la Turquie. Rev. Scientif., Paris 1947, 716–724.
Steffens, P.: Remarks on the Upper Cenozoic of West Anatolia. Newsl. Stratigr., Leiden, 1 (3) (1971), 47–49.
Stöcklin, J.: Structural history and tectonics of Iran. Bull. Am. Ass. Petr. Geol., Tulsa, 52 (1968a), 1229–1256.
Stöcklin, J.: Salt deposits of the Middle East. In: *Mattox, R. B.* (Ed.): Saline deposits. Geol. Soc. Am. Spec. Pap., Boulder, 88 (1968b), 157–181.
Stöcklin, J., A. Ruttner and *M. Nabavi:* New data on the Lower Paleozoic and Pre-Cambrian of North Iran. Rep. Geol. Surv. Iran, Teheran, 1 (1964), 29 p.
Strid, A. (Ed.): Evolution in the Aegean. Opera Bot., Lund, 30 (1971), 83 p.
Striebel, H.: Die Bleierz-Baryt-Lagerstätte von Karalar/Gazipasa (Türkei) und ihr geologischer Rahmen. Diss. Naturwiss. Fak. Univ. München, München 1965, 48 p.

Taşman, C. E.: Salt domes of Central Anatolia. MTA Mecm. 2 (4) (1937), 43–46.
Taşman, C. E.: The stratigraphy of the Alexandrette Gulf basin. C. R. 18. Congr. Géol. Int., London, pt. 6 (1950), 65–67.
Taşman, M.: Foraminifera from test wells in Adana, Turkey. Publ. MTA B 15 (1949), 42 p.
Tatar, Y.: Ofiolitli Camlıbel dolaylarında jeolojik ve petrografik araştırmalar (Yıldızeli/Iç Anadolu). Doc. Tezi Karadeniz Tekn. Üniv., Trabzon 1971, 167 p.
Tchalenko, J. S., J. Braud and *M. Berberian:* Discovery of three earthquake faults in Iran. Nat., London, 248 (1974), 661–663.
Tchernisheva, N. E.: Middle Cambrian trilobites of the northern Caucasus. Paleont. Journ., Moskva 1968 (1), 71–80.
Tenchov, Y. G.: The Carboniferous System in Bulgaria. C. R. 6. Int. Congr. Stratigr. Géol. Carbonifère, Maastricht, 4 (1971), 1543–1553.
Ten Dam, A.: Sedimentation, facies and stratigraphy in the Neogene basin of Iskenderun. Bull. GST 3 (2) (1952), 49–64.
Ten Dam, A.: Stratigraphy and sedimentation of the Lower Tertiary and Mesozoic in the foredeep basin of SE-Turkey. Bull. GST 6 (1) (1955), 135–155.
Ten Dam, A.: La bordure nord de la plate-forme arabique. C. R. Soc. Géol. Fr., Paris 1965, 153–156.
Ten Dam, A. and *N. Tolun:* Struttura e geologia della Turchia. Bull. Soc. Geol. It., Roma, 80 (1962), 45–80.
Ternek, Z.: Mersin Tarsus kuzey bölgesinin jeolojisi. MTA Mecm. 18 (44/45) (1953a), 18–62.
Ternek, Z.: Geological study of southeastern region of Lake Van. Bull. GST 4 (2) (1953b), 1–32.
Ternek, Z.: The Lower Miocene (Burdigalien) formations of the Adana basin, their relations with other formations, and oil possibilities. Bull. MTA 49 (1957), 60–80.

Ternek, Z.: A geological note on the natural gas in Söke. Bull. GST 7 (1) (1959), 58–74.
Thiele, O.: Der Nachweis einer intrapermischen Faltungsphase im westlichen Zentral-Iran. Verh. Geol. Bundesanst., Wien 1973, 489–498.
Thompson, G. and *W. G. Melson:* The petrology of oceanic crust across fracture zones in the Atlantic Ocean: evidence of a new kind of sea-floor spreading. Journ. Geol., Chicago, 80 (1972), 526–538.
Tidewater Oil Co.: Geology of the Sinop area. Petrol. Activ., Ankara, 6 (1961), 33–34.
Tietze, E.: Beiträge zur Geologie von Lykien. Jahrb. Geol. Reichsanst., Wien, 35 (1885), 283–386.
Tilev, N.: Étude des Rosalines maestrichtiennes (genre Globotruncana) du Sud-Est de la Turquie (sondage de Ramandağ). Publ. MTA B 16 (1951), 101 p.
Tokay, M.: Contribution à l'étude de la région comprise entre Ereğli, Alaplı, Kızıltepe et Alacaağzı. MTA Mecm. 42/43 (1952), 37–78.
Tokay, M.: Géologie de la région de Bartın (Zonguldak, Turquie de Nord). Bull. MTA 46/47 (1954/55), 46–63.
Tokay, M.: The geology of the Amasra region with special reference to some Carboniferous gravitational gliding phenomena. Bull. MTA 58 (1962), 1–20.
Tollmann, A.: Das Strandscha-Fenster, ein neues Fenster der Metamorphiden im alpinen Nordstamm des Balkans. N. Jahrb. Geol. Pal. Monatsh., Stuttgart 1965, 234–248.
Tolun, N.: Notes géologiques sur la région de Silvan–Hazru. Bull. GST 2 (1) (1949), 65–89.
Tolun, N.: Étude géologique du bassin nord-est de Diyarbakır. MTA Mecm. 16 (41) (1951), 65–98.
Tolun, N.: Contribution à l'étude des environs du S et SW du Lac de Van. MTA Mecm. 44/45 (1953), 77–114.
Tolun, N.: Stratigraphy and tectonics of southeastern Anatolia. Rev. Ist. 25 (1962), 203–264.
Tolun, N. and *Z. Ternek:* Notes géologiques sur la région de Mardin. Bull. GST 3 (2) (1952), 1–16.
Toula, F.: Eine Muschelkalkfauna am Golfe von Ismid in Kleinasien. Beitr. Pal. Österr.-Ung., Wien, 10 (1896), 153–191.
Toula, F.: Übersicht über die geologische Literatur der Balkanhalbinsel mit Morea, des Archipels mit Kreta und Cypern, der Halbinsel Anatolien, Syrien und Palästinas. C. R. 9. Congr. Géol. Int., Wien, 1 (1904), 185–330.
Türkünal, M.: Notes on some Lower Cretaceous ammonites from Karalar Köyü, northwest of Ankara. Bull. MTA 50 (1958), 75–79.
Türkünal, M.: Note on the ammonite-bearing beds in the various localities of Turkey. Part one: Ankara region. Bull. MTA 52 (1959), 67–74.
Türkünal, S.: Note sur la géologie des montagnes de Hakkâri. Bull. GST 3 (1) (1951), 33–43.
Türkünal, S.: Géologie de la région de Hakkâri et de Başkale (Turquie). Publ. MTA B 18 (1953), 43 p.
Türkünal, S.: Contribution à l'étude géologique de la région située entre Çukurca, Beytüşşebap et Şirnak. Bull. GST 6 (1) (1955), 50–60.
Turdok: Monthly current awareness service on pure and applied scientific articles, ser. B (foreign ed.). Ankara.
Turkish Golf Oil Co.: Regional geology and oil possibilities in the Tuz Gölü basin of Central Anatolia. Petrol. Activ., Ankara, 6 (1961), 29–32.
Tuscu, N.: Étude minéralogique et pétrographique de la région de Başkışla (Karaman–Konya, Taurus occidental, Turquie). Thèse Univ. Genève, Genève 1972, 106 p.

Udluft, H.: Untersuchungen an Eruptivgesteinen aus der Umgegend von Konstantinopel. Abh. Preuss. Geol. Landesanst. N. S., Berlin, 190 (1939), 1–19.
Ünsalaner, C.: A preliminary description of the Carboniferous and Devonian fauna discovered in the western Taurus. MTA Mecm. 6 (4/25) (1941), 594–603.
Ünsalaner, C.: Some Upper Devonian corals and stromatoporids from South Anatolia. Bull. GST 3 (1) (1951), 131–144.
Ünsalaner-Kırağlı, C.: Lower Carboniferous corals from Turkey. Journ. Pal. Soc. India, Lucknow, 3 (1958), 53–58.
Uz, B.: Les formations métamorphiques et granitiques du massif ancien d'Akdağ (Simav, Turquie) et leur couverture volcano-sédimentaire. Thèse Univ. Nancy I, Sciences Terre, Nancy 1973, 331 p.
Uzkut, I.: Die Manganerzvorkommen von Türkisch-Thrakien. Diss. Univ. Clausthal, Clausthal 1971, 326 p.

Vaché, R.: Die Kontaktlagerstätte von Akdağmadeni und ihr geologischer Rahmen innerhalb des zentral-anatolischen Kristallins. Bull. MTA 60 (1963), 22–36.
Vaché, R.: Die Blei-Zinkerzlagerstätte am Bakırdağ im Antitaurus (Prov. Kayseri). Bull. MTA 62 (1964), 91–102.
Vachette, M., Ph. Blanc and *L. Dubertret:* Détermination de l'âge d'une granodiorite d'Orhaneli au sud de Bursa (Anatolie); sa signification régionale. C. R. Ac. Sc., Paris, D 267 (1968), 927–930.

Van Andel, T.: Recent marine sediments of Gulf of California. In: *Van Andel, T.* and *G. G. Shor* (Eds.): Marine geology of the Gulf of California. Am. Ass. Petrol. Geol. Mem., Tulsa, 3 (1964), 216–310.
Van Der Kaaden, G.: Age relations of magmatic activity and metamorphic processes in the northwestern part of Turkey. Bull. MTA 52 (1959), 15–33.
Van Der Kaaden, G.: The significance and distribution of glaucophane rocks in Turkey. Bull. MTA 67 (1966), 36–67.
Van Der Kaaden, G.: Zur Entstehung der Glaukophan-Lawsonit- und glaukophanitischen Grünschiefer-Fazies. Geländebeobachtungen und Mineralsynthesen. Fortschr. Miner., Stuttgart, 46 (1969), 87–136.
Van Der Kaaden, G.: Chromite-bearing ultramafic and related gabbroic rocks and their relationship to „ophiolitic" extrusive rocks and diabases in Turkey. Geol. Soc. South Afr. Spec. Publ., Pretoria, 1 (1970), 511–531.
Van Der Kaaden, G.: Basement rocks of Turkey. In: *Campbell, A. S.* (Ed.): Geology and History of Turkey. Tripoli 1971, p. 191–209.
Van Wijkerslooth, P.: Über den jungen Vulkanismus am Innenrande des Taurus zwischen Afyon–Karahisar und Kayseri (Türkei). MTA Mecm. 9 (2/32) (1944), 244–256.
Van Wijkerslooth, P.: Über das Alter und die Genese der Kupfer-Erzlagerstätte „Ergani Maden" (Vilayet Elaziğ, Türkei). Bull. GST 5 (1954), 190–198.
Van Wijkerslooth, P. and *H. Kleinsorge*: Zur Geologie der devonischen oolithischen Eisenerzlagerstätte Çamdağ bei Adapazarı, Vilayet Kocaeli (Izmit), Türkei. MTA Mecm. 5 (3/20) (1940), 319–334.
Van Zeist, W.: Paleobotanical results of the 1970 season at Çayönü, Turkey. Helinium, Wetteren, Belg., 12 (1972), 3–19.
Verdier, J.: Étude du Kemalpaşa Dağı (province d'Izmir, Turquie). Bull. MTA 61 (1963), 38–40.
Vinogradov, A. P., A. B. Ronov and *V. E. Khain*: Atlas of litho-paleogeographical maps of the Russian Platform and its geosynclinal frame. Pt. 2, Mesozoic and Cenozoic. 95 maps 1:500 000. Moskva–Leningrad 1961.
Vogt, P. R. and *R. H. Higgs*: An aeromagnetic survey of the Eastern Mediterranean Sea and its interpretation. Earth Planet. Sc. Letter, Amsterdam, 5 (1969), 439–448.
Vohryzka, K.: Zur Geologie und Metallogenese des Gebietes zwischen Yahyalı (Kayseri) und Zamantıfluß. Bull. MTA 67 (1966), 97–104.
Von Arthaber, G.: Die Trias von Bithynien (Anatolien). Beitr. Pal. Österr.-Ung. Orient, Wien, 27 (1915), 85–206.
Von Bukowski, G.: Die geologischen Verhältnisse der Umgebung von Balia Maden im nordwestlichen Kleinasien (Mysien). Sitzber. Ak. Wiss. Math.-nat. Kl. pt. I, Wien, 101 (1892), 214–235.
Von Bukowski, G.: Neuere Fortschritte in der Kenntnis der Stratigraphie von Kleinasien. C. R. 9. Congr. Géol. Int., Wien, 1 (1904), 393–426.
Von Gaertner, H.-R.: Zur tektonischen und magmatischen Entwicklung der Kratone. Beih. Geol. Jahrb., Hannover, 80 (1969), 117–145.
Vuagnat, M. and *E. Çoğulu*: Quelques réflexions sur le massif basique-ultrabasique du Kızıl Dagh, Hatay, Turquie. C. R. Séanc. Soc. Phys. Hist. Nat. N. S., Genève, 2 (1968), 210–216.

Wagner, R. H.: On a mixed Cathaysia and Gondwana flora from SE-Anatolia (Turkey). C. R. 4. Congr. Int. Stratigr. Géol. Carbonifère, Maastricht, 3 (1962), 745–752.
Wagner-Gentis, C. H. T.: Upper Visean goniatites from northern Anatolia. Bull. MTA 50 (1958), 80–82.
Weber, H.: Ergebnisse erdölgeologischer Aufschlußarbeiten der DEA in Nordost-Syrien. Erdöl u. Kohle, Hamburg (1963/64), 16, 669–682, 17, 249–261.
Wedding, H.: Beiträge zur Geologie der Kelkitlinie und zur Stratigraphie des Jura im Gebiet Kelkit–Bayburt. Bull. MTA 61 (1963), 31–37.
Wedding, H.: Über eine interessante Blattverschiebung ostwärts Bartın. Bull. MTA 74 (1970), 43–51.
Westerveld, J.: Phases of Neogene and Quaternary volcanism in Asia Minor. C. R. 20. Int. Geol. Congr. pt. 1 (1), Mexico (1957), 103–119.
Wettstein, O.: Herpetologia aegaea. Sitzber. Ak. Wiss. Math.-nat. Kl. pt. I, Wien, 162 (1953), 651–833.
Wiesner, K.: Konya civa yatakları ve bunlar üzerindeki etüdler. MTA Derg. 70 (1968), 178–213.
Wijmstra, T. A.: The place of the Tenagi Philippon in the Pleistocene stratigraphical sequence. Zeitschr. Deutsch. Geol. Ges., Hannover, 123 (1972), 565–566.
Wilser, J. L.: Die stratigraphische und tektonische Stellung der Dobrudscha und die Zugehörigkeit des Balkangebirges zu den nordanatolischen Ketten. Geol. Rundsch., Berlin, 19 (1928), 161–223.
Wilson, J. T.: A new class of faults and their bearing on continental drift. Nat., London, 207 (1965), 343–347.
Winkler, H. G. F.: Die Genese der metamorphen Gesteine. Berlin–Heidelberg–New York 1967, 2nd ed., 237 p.
Wippern, J.: Die Bauxite des Taurus und ihre tektonische Stellung. Bull. MTA 58 (1962), 47–70.

Wippern, J.: Die Stellung des Menderes-Massivs in der alpidischen Gebirgsbildung. Bull. MTA 62 (1964a), 74–82.
Wippern, J.: Die Aluminium-Rohstoffe der Türkei. Bull. MTA 62 (1964b), 83–90.
Wippern, J.: Die Ausgangsgesteine für die Bauxitbildung. Bull. MTA 64 (1965), 40–44.
Wirtz, D.: Beobachtungen zur jüngeren geologischen Entwicklung Anatoliens (Beispiele aus der Ägäis und Zentralanatolien). Geol. Jahrb., Hannover, 76 (1958), 325–334.
Wolfart, R.: Geologie von Syrien und dem Libanon. Beitr. Reg. Geol. d. Erde, Berlin, 6 (1967a), 326 p.
Wolfart, R.: Zur Entwicklung der paläozoischen Tethys in Vorderasien. Erdöl und Kohle, Hannover, 20 (1967b), 168–180.
Wolfart, R. and *M. Kürsten:* Stratigraphie und Paläogeographie des Kambriums im mittleren Süd-Asien (Iran bis Nord-Indien). Geol. Jahrb., Hannover, B 8 (1974), 185–234.
Wong, H. K., E. F. K. Zarudzki, J. D. Phillips and *G. K. F. Giermann:* Some geophysical profiles in the eastern Mediterranean. Bull. Geol. Soc. Am., Boulder, 82 (1971), 91–99.
Woodside, J. and *C. Bowin:* Gravity anomalies and inferred crustal structure in the eastern Mediterranean Sea. Bull. Geol. Soc. Am., Boulder, 81 (1970), 1107–1122.

Yahşıman, K.: Palynology and correlation in the Zonguldak coal basin. Rev. Ist. 37 (1972), 249–264.
Yalcınlar, I.: Nouvelles observations sur les terrains paléozoiques des environs d'Istanbul. Bull. GST 3 (1) (1951), 127–130.
Yalcinlar, I.: Istanbulda bulunan graptolitli Silür sistleri hakkında. Ist. Üniv. Çogr. Enst. Derg., Istanbul, 4 (1956), 157–160.
Yalcınlar, I.: Samsun bölgesinin Neojen ve Kuaterner kiyi depolari. Ist. Üniv. Çogr. Enst. Derg., Istanbul, 5 (9) (1958), 12–21.
Yalcınlar, I.: Sur le terrain du Primaire ancien au sud d'Akşehir (Turquie). C. R. Soc. Géol. Fr., Paris 1959, 155–156.
Yalcınlar, I.: Le massif calédonien de Babadağ et ses couvertures anthracolithiques. Bull. MTA 60 (1963), 14–21.
Yalcınlar, I.: Observations sur la faune du Primaire ancien dans la région méditerranéenne de la Turquie. Bull. GST 16 (1) (1973), 101–109.
Yazgan, E.: Étude géologique et pétrographique du complexe ophiolitique de la région située au sud-est de Malatya (Taurus oriental, Turquie) et de sa couverture volcano-sédimentaire. Thèse Fac. Sc. Univ. Genève, Genève 1972, 236 p.
Yazlak, Ö.: Sur quelques Pectinidae de Turquie. Bull. GST 16 (1) (1973), 110–131.
Yılmaz, I.: Étude pétrogénétique des granites d'Alacam Dağları. Abstract. Congr. Earth Sc., Ankara 1973, p. 61.
Yılmaz, O.: Étude pétrographique et géochronologique de la région de Cacas (partie méridionale du massif de Bitlis, Turquie). Thèse Fac. Sc. Terre Univ. Grenoble, Grenoble 1971, 230 p.
Yüksel, S.: Étude géologique de la région de Haymana (Turquie Centrale). Thèse Fac. Sc. Univ. Nancy, Nancy 1970, 179 p.
Yurttaş-Özdemir, Ü.: Kocaeli yarımadası, Tepeköy triası makrofaunası ve biyostratigrafisi. MTA Derg. 77 (1971), 57–98.
Yurttaş-Özdemir, Ü.: Über den Schieferton mit Halobia der Halbinsel Kocaeli. Bull. MTA 80 (1973), 43–49.

Zapfe, H.: Fragen und Befunde von allgemeiner Bedeutung für die Biostratigraphie der alpinen Obertrias. Verh. Geol. Bundesanst., Wien 1967, 13–27.
Zeiller, R.: Étude sur la flore fossile du bassin houiller d'Héraclée. Mém. Soc. Géol. Fr., Paris, 21 (1899), 91 p.
Zijderveld, J. D. A. and *R. Van Der Voo:* Paleomagnetism in the Mediterranean area. In: *Tarling, D. H.* and *S. K. Runcorn* (Eds.): Implications of continental drift to the Earth Sciences, London–New York 1973, p. 133–161.
Zijlstra, G.: Erosion of the Namurian during the Westphalian B–C in the Zonguldak coal field (Turkey). MTA Mecm. 42/43 (1952), 121–122.
Zwittkovits, F.: Klimabedingte Karstformen in den Alpen, den Dinariden und im Taurus. Mitt. Österr. Geogr. Ges., Wien, 108 (1966), 72–97.

Index of geographic and regional geologic names